Dark Matter in the Universe

Dark Matter in the Universe

Marc S Seigar

Department of Physics and Astronomy,
University of Minnesota Duluth, USA

Morgan & Claypool Publishers

ISBN   978-1-6817-4118-5 (ebook)
ISBN   978-1-6817-4054-6 (print)
ISBN   978-1-6817-4246-5 (mobi)

DOI   10.1088/978-1-6817-4118-5

Version: 20150901

IOP Concise Physics
ISSN 2053-2571 (online)
ISSN 2054-7307 (print)

A Morgan & Claypool publication as part of IOP Concise Physics
Published by Morgan & Claypool Publishers, 40 Oak Drive, San Rafael, CA, 94903, USA

IOP Publishing, Temple Circus, Temple Way, Bristol BS1 6HG, UK

For my wife, Colleen, and my children, David and Andrew

Contents

Preface

The study of dark matter encompasses three main areas in fundamental physics: astrophysics, cosmology, and particle physics. As such, it is difficult to cover every aspect of dark matter in a concise book, such as this, and so this book is intended as an introduction for beginning physics majors, or those interested in a short course in dark matter.

This book starts by giving a brief historical overview of why dark matter is a necessary concept in modern physics, at least from an astrophysical perspective. The first three chapters focus on the astrophysical necessity for dark matter, and why it is necessary if we want to be able to describe the structures that we see in the Universe, particularly on the largest scales. The next three chapters focus on the particle physics necessary to understand dark matter. I have chosen to focus on just a few possible forms of proposed dark matter: Weakly Interacting Massive Particles or WIMPs (whether predicted by supersymmetric or non-supersymmetric models), Super Weakly Interacting Massive Particles or SuperWIMPs, sterile neutrinos, and axions. A brief overview of the Standard Model of particle physics is given along with the need for extensions to it. All of these extensions predict the existence of new particles, some of which have predicted characteristics that would be necessary for dark matter. The final chapter summarizes our modern cosmological model and how dark matter fits in. The final chapter also includes some possible scenarios that may play out within the next decade or so.

I hope that the readers of this book are inspired to learn more about the subject of dark matter.

Marc S Seigar
Department of Physics and Astronomy
University of Minnesota Duluth
Duluth, MN, USA
5 September 2015

Acknowledgements

I wish to thank all of my family and friends. In particular, I wish to thank my PhD advisor, Professor Philip James, for putting up with me in Graduate School. I wish to thank all of my collaborators, in particular Andy Adamson, Misty Bentz, Joel Berrier, David Block, Jane Buckle, James Bullock, David Buote, Tim Carroll, Malcolm Currie, John Davies, Benjamin Davis, Chris Davis, Herwig Dejonghe, Tim de Zeeuw, Fabio Gastaldello, Alister Graham, Philip Humphrey, Helmut Jerjen, Daniel Kennefick, Julia Kennefick, Tom Kerr, Olga Kuhn, Claud Lacy, Sandy Leggett, Burcin Mutlu, Ivanio Puerari, Nicholas Rees, Heikki Salo, Doug Shields, Amber Sierra, Massimo Stiavelli, Patrick Treuthardt, Thor Wold, Watson Varricatt, and Luca Zappacosta. If I did not remember you, I apologize. I have had many collaborators over the years.

I want to thank all of my friends and colleagues who have helped me through the years, particularly Haydar Al-Shukri, Matt Andrews, Ann Bain, Brian Berry, John Bush, Chris Collins, Toni Empl, Jeff Gaffney, Micheal Gealt, Anindya Ghosh, Joshua Hamilton, Keith Hudson, Darin Jones, Tansel Karabacak, Johanna Lewis, Howard Mooers, Patrick Pellicane, Julian Post, Jeff Robertson, Jim Rock, Derek and Hazel Sears, and Amber Straughn.

My friends whom I have had so much fun over the years, I thank you all: Liz Alvarez, Roberto Avila, Renee and Kirt Booth, Alex Bouquin, Marc Cohen, Richard Dolman, Michael Fahrenwald, Paul Harman, Breanna and Bill Johnson, Melissa and Ivan Mitchell, Erik Rau and Elle Bublitz, Helene Schuller and Todd Chelson, Jamie and Dan Sweeney, Sonali and Gavin Whitlock, amongst others.

I wish to thank the editorial and productions teams at Morgan & Claypool and IOP Publishing for making the process of writing and publishing this book fairly easy and straightforward.

Most importantly, I would like to thank my parents, Vivienne and Michael Seigar, who always encouraged me. Without them, I would not be where I am today.

Author biography

Marc S Seigar

Marc S Seigar is a Professor of Physics and Astronomy and the Head of the Department of Physics and Astronomy at the University of Minnesota Duluth (UMD). He is also the Director of the Marshal W Alworth Planetarium at UMD. Prior to his arrival at UMD, he worked as a Professor of Astrophysics at the University of Arkansas at Little Rock, a Project Scientist at the University of California, Irvine, and a Staff Astronomer at the United Kingdom Infrared Telescope (UKIRT). Professor Seigar has published numerous papers and conference proceedings articles in the field of galaxy dynamics, spiral structure, and dark matter.

Chapter 1

The need for dark matter

Figure 1.1. Fritz Zwicky. Image courtesy of scienceblogs.com.

Fritz Zwicky (a photo of whom can be seen in figure 1.1) was born in Bulgaria, a citizen of Switzerland, and did most of his best work at the California Institute of Technology in Pasadena, California. He was in the very tough spot of having a number of great and correct ideas that, by and large, people did not take seriously. One of them was the discovery of 'missing mass' in the Universe.

While examining the Coma cluster of galaxies in 1933, Fritz Zwicky was the first to infer the existence of unseen matter. He coined the phrase *dunkle materie* or dark matter. Using the Newtonian law of gravity, Zwicky calculated the gravitational mass of the galaxies within the Coma cluster and obtained a value that was considerably higher (at least 400 times higher) than that expected from the starlight being emitted by all of the galaxies in the cluster. This meant that most of the matter in the cluster was unseen or dark, thus the term dark matter. While Zwicky had overestimated the amount of dark matter in the Coma cluster, the same calculation based on better, more recent data, still indicates that the majority of matter in galaxy clusters appears to be dark.

The Coma Cluster, as shown in figure 1.2 in an image from NASA's Hubble Space Telescope, is a collection of thousands of galaxies that are gravitationally bound into a single, spherical volume more than 20 million light-years in diameter. This Hubble Space Telescope image captures the magnificent starry population of the Coma Cluster of galaxies, one of the densest known galaxy collections in the Universe. However, the visible galaxies are just a small part of the picture as most of the space in this cluster is filled with invisible, dark matter. The need for dark matter, while originally discovered by Fritz Zwicky, was never really taken seriously until Vera Rubin confirmed the result by looking at individual galaxies four decades later.

Figure 1.2. The Coma Cluster. Image credit: NASA, ESA, and the Hubble Heritage Team (STScI/AURA). Acknowledgement: D Carter (Liverpool John Moores University) and the Coma Hubble Space Telecope Treasury Team.

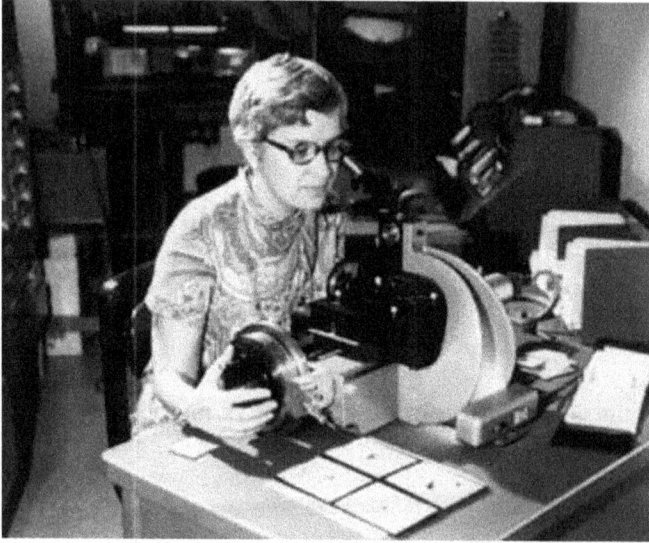

Figure 1.3. Vera Rubin. Image courtesy of the American Museum of Natural History.

In the 1970s, Rubin (pictured in figure 1.3) began research on the rotation curves of galaxies, starting with the Andromeda Galaxy. A rotation curve is a plot of orbital velocity versus distance from the galactic center, and this is what Rubin studied for several galaxies from the early 1970s through to the mid-1980s. She pioneered this field by showing that material in galaxies (the stars and gas) was moving too fast. Galaxies would fly apart if the material holding them together gravitationally was just made up of their constituent stars. However, galaxies do not fly apart, and therefore a huge amount of unseen mass (or dark matter) must be holding them together. Rubin's calculations showed that galaxies must contain at least ten times as much dark mass as can be accounted for by the visible stars and gas. Attempts to explain this discrepancy led to the theory of dark matter being much more widely accepted, but this may have been partly due to research that was happening in parallel, within a different field of physics.

In the 1970s and 1980s, in particle physics, it was realized that the Standard Model of particle physics had some gaps and some aspects that simply could not be explained. For instance, it does not include gravity, but it does include the other three fundamental forces of nature (electromagnetism, the weak nuclear force, and the strong nuclear force). Also, gravity is an incredibly weak force, and there can be huge differences in mass between different fundamental particles, which is not adequately explained by the Higgs mechanism. This leads to new theories, all of which are essentially extensions to the Standard Model of particle physics. The most popular of these extensions is known as supersymmetry. In theories of supersymmetry, every Standard Model particle has a supersymmetric partner. This predicts that more fundamental particles should exist, and this can explain the discrepancies between the strengths of fundamental forces and the differences between the masses of the fundamental particles. Some of these 'new' particles have some intriguing properties. Supersymmetry (as well as some other models) predicts a class of particle known as weakly interacting massive particles or WIMPs. These particles have high masses, but they have a very small chance of ever interacting electromagnetically or interacting through the strong nuclear force. Due to their large masses, however, they can have a strong gravitational influence. Suddenly, there was a theory that had a prediction of what dark matter could be.

Our particle physics models now predict a host of particles that could potentially be candidates for dark matter (see chapter 4 for more details). They all have little or no electromagnetic interaction or strong nuclear interaction. One could imagine that the gauge bosons that carry these forces (photons and W and Z bosons) essentially pass right by these particles without being affected. In other words, these dark matter candidates act as if they are transparent to photons. If we were naming this type of matter today, we might call it transparent matter rather than dark matter. Indeed, transparent matter is a much better description of the underlying physical processes than dark matter.

In a nutshell, this introductory chapter has outlined the dark matter problem. In this book you will be able to read all about it. We will start by discussing how dark matter is detected through its gravitational effect in galaxies and clusters of galaxies, and how our theories explain this. We will talk about the different types of proposed dark matter particles. We will learn about how dark matter can potentially be detected, and about the experiments scientists are building to detect dark matter. We will even discuss what it means if we never detect dark matter, other than through its gravitational interaction with normal matter. This book is written for the undergraduate student with an interest in frontier physics. It is aimed as an introduction to the topic of dark matter cosmology. I think it is an intriguing story, and I hope you do too.

Suggested further reading

Rubin V C and Ford W K Jr 1970 *Astrophys. J.* **159** 379–403
Rubin V C, Thonnard N and Ford W K Jr 1978 *Astrophys. J.* **225** L107–10
Rubin V C, Ford W K Jr and Thonnard N 1980 *Astrophys. J.* **238** 471–87
Rubin V C, Ford W K Jr, Thonnard N and Burstein D 1982 *Astrophys. J.* **261** 439–56
Rubin V C, Burstein D, Ford W K Jr and Thonnard N 1985 *Astrophys. J.* **289** 81–98, 101–4
Zwicky F 1933 *Helv. Phys. Acta* **6** 110–27

Chapter 2

The formation of structure and dark matter in Galaxies

In this chapter, we will discuss the need for dark matter from an astrophysical point of view. We will see that dark matter is needed not only to explain the dynamics of galaxies, but also so that we can explain the observed large-scale structure in the Universe. This will unveil why dark matter is a cornerstone in the study of modern cosmology.

The motion of stars and gas in spiral galaxies provides a means of measuring the mass of such galaxies simply using the ideas of Newtonian gravity. From Newtonian gravity, we know that the velocities of stars around a galaxy follow a simple relation given by,

$$v_{\text{rot}} = \sqrt{\frac{GM}{r}} \qquad (2.1)$$

where r is the distance from the center of the galaxy, M is the mass contained within the distance r, and $G = 6.67 \times 10^{-11} \, \text{m}^3 \, \text{kg}^{-1} \, \text{s}^{-2}$ is Newton's gravitational constant. This tells us that the rotational velocity, v_{rot}, falls off inversely as the square root of the distance from the center of the galaxy.

In the late 1960s, Vera Rubin started to take observations of the rotation velocities of spiral (or disk) galaxies (otherwise known as rotation curves). One of the most famous rotation curves she obtained was that of M31, the Andromeda Galaxy (see figure 2.1). The main result of this study was that the rotation velocities of material in the Andromeda Galaxy remain high at very large galactocentric radii. However, according to the equation for the rotational velocities above, this was unexpected. From the visible mass in the galaxy (i.e., the stars), we would expect the velocities to fall off as the inverse of the square root of the distance. The only way we can reconcile this observational result with our theories of gravity is if the enclosed mass, M, is higher than observed (about 4–5 times higher), and also if the dark

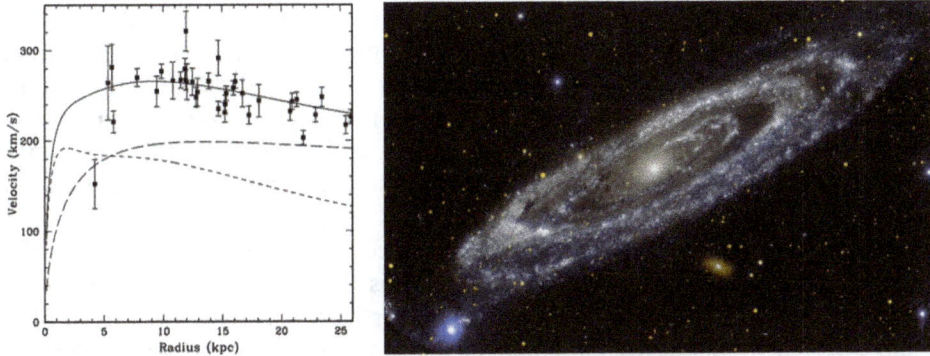

Figure 2.1. The rotation curve of the Andromeda galaxy (left) and an ultraviolet image of the Andromeda galaxy (right). Rotation curve credit (left): from Seigar *et al* (2008, Monthly Notices of the Royal Astronomical Society, Vol. 208, pp 1911–23). Image credit (right): NASA Galaxy Evolution Explorer Team.

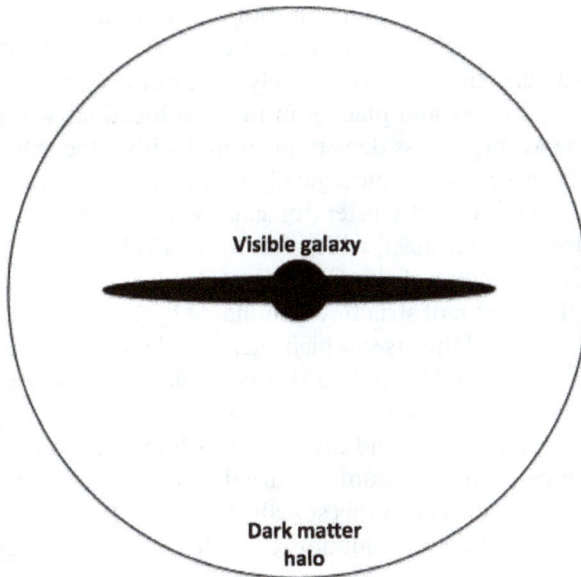

Figure 2.2. A schematic diagram of a visible galaxy in a dark matter halo (not to scale). The dark matter halo is typically 5–6 times the radius of the visible galaxy at its center.

matter extends well beyond the visible part of the galaxy. Therefore, the conclusion from these rotation curve data is that galaxies must contain much more mass than the visible light would otherwise indicate. Indeed, our modern picture of galaxies has a visible galaxy sitting at the center of a sphere of dark matter with a radius that is 5–6 times larger than that of the visible galaxy. A schematic diagram of this is shown in figure 2.2.

These so-called 'flat' rotation curves were confirmed in a series of investigations in the late 1970s and early 1980s. These rotation curves have now been extended to larger radii with sensitive measurements of neutral hydrogen gas, demonstrating that the rotation velocities remain high far beyond the visible disk in galaxies. This suggests that the visible light (and thus stellar mass) accounts for only a small fraction (typically 15% or smaller) of the mass in spiral galaxies. This still remains one of the best pieces of evidence in favor of dark matter cosmology. In order to get the flat rotation curves we observe, the simplest model is such that the dark matter exists in a large extended (almost) spherical halo surrounding the visible galaxy (see figure 2.2).

It has now been more than 40 years since the discovery that galaxies are surrounded by extended massive halos of dark matter. In this time, a variety of observational probes have made it possible to map dark matter halo mass distributions in some detail. These distributions are intimately linked to the nature of the dark matter, the way dark halos formed, and the cosmological context of dark halo formation.

The basic assumptions of modern cosmology are that the Universe is homogeneous and isotropic (the so-called cosmological principle). Homogeneous means that the Universe looks the same from every point within it. Isotropic means that the Universe looks the same in every direction. Both of these taken together means that there is no special place in the Universe, and the Universe has no center. On the largest scales, this assumption works extremely well. For instance, if we take a sphere of radius 4000 mega-parsecs and place it in random locations within the Universe, the variation in the average mass density measured within the sphere will be about 1 part in 10 000. However, on cosmologically small scales (i.e., galaxy-sized scales), just the fact that galaxies (and clusters of galaxies) exist, suggests that there are inhomogeneities (or overdensities), which cannot be explained by the cosmological principle alone. The structure of the Universe becomes much more complicated at these scales, with the growth of structure dominated by gravity. The standard model for the cosmology of the Universe (which includes dark matter and dark energy) has proven to be an invaluable tool as this cosmology reproduces the large-scale structures observed in the Universe extremely well.

In the tiniest fraction of a second after the Big Bang (up to about 10^{-32} s), the Universe was dominated by quantum perturbations or fluctuations. Particles were essentially created out of energy in this so-called hot Big Bang model. This is one of the results of Einstein's famous equation that showed that energy and mass are related to each other, i.e.,

$$E = mc^2 \tag{2.2}$$

where E is energy, m is rest mass, and $c = 2.998 \times 10^8 \, \mathrm{m \, s^{-1}}$ is the speed of light. Equation (2.2) is often referred to as mass-energy equivalence. In other words, mass is simply another form of energy. Particles (or quantum fluctuations) being created in the hot Big Bang model are also a result of the Heisenberg uncertainty principle, which can be written as

$$\sigma_x \sigma_p \geqslant \frac{h}{4\pi} \tag{2.3}$$

where σ_x is the uncertainty in a particle's measured position, σ_p is the uncertainty in its momentum, and $h = 6.626 \times 10^{-34}$ J s is Planck's constant. Equation (2.3) means that pairs of particles can be created and destroyed all the time, if the two uncertainties in equation (2.3), taken together, violate the condition of the equation. If the condition is violated, it is as if the particles do not exist, and they are referred to as virtual particles.

Under some conditions, however, it is possible to transform virtual particles into real particles. For this, we go back to equation (2.2), where even a tiny amount of mass can be equivalent (or can be converted to) a lot of energy. Furthermore, if there is enough energy available, particles can be created with the equivalent amount of mass. As mentioned above, mass is simply another form of energy. The particles are always created in pairs, a particle and an antiparticle. Under most circumstances, they would annihilate each other and their mass would be converted back into energy. However, at these early times, the Universe underwent an inflationary expansion, which took place between 10^{-36} s and 10^{-32} s after the Big Bang (the Big Bang being defined as the very beginning when time $t = 0$). During this brief amount of time, the Universe expanded extremely rapidly, at an exponentially increasing rate. Why do we believe this happened?

An expanding Universe generally has a cosmological horizon, which, by analogy with the more familiar horizon caused by the curvature of the Earth's surface, marks the boundary of the part of the Universe that an observer can see. Light (or other radiation) emitted by objects beyond the cosmological horizon never reaches the observer, because the space in between the observer and the object is expanding too rapidly. The observable Universe is one causal patch of a much larger unobservable Universe; there are parts of the Universe that cannot communicate with us yet. These parts of the Universe are outside our current cosmological horizon. In the standard hot Big Bang model, without inflation, the cosmological horizon moves out, bringing new regions into view. Yet as a local observer sees these regions for the first time, they look no different from any other region of space the local observer has already seen: they have a background radiation that is at nearly exactly the same temperature as the background radiation of other regions, and their space–time curvature is evolving lock-step with ours. This presents a mystery: how did these new regions know what temperature and curvature they were supposed to have? They couldn't have learned it by getting signals, because they were not in communication with our past light cone before.

Inflation answers this question by postulating that all the regions come from an earlier era with a big vacuum energy, or cosmological constant. A space with a cosmological constant is qualitatively different: instead of moving outward, the cosmological horizon stays put. For any one observer, the distance to the cosmo-logical horizon is constant. With exponentially expanding space, two nearby observers are separated very quickly; so much so, that the distance between them quickly exceeds the limits of communications. The spatial slices are expanding very fast to cover huge volumes. Things are constantly moving beyond the cosmological horizon, which is a fixed distance away, and everything becomes homogeneous very quickly. As the inflationary field slowly relaxes to the vacuum, the cosmological constant goes to zero, and space begins to expand normally. The new regions that come into view during the normal expansion phase are exactly the same regions that were pushed out of the horizon during inflation, and so they are necessarily at nearly the same temperature and curvature, because they come from the same little patch of space.

The inflationary model of the Universe thus explains why the temperatures and curvatures of different regions are so nearly equal. It also predicts that the total curvature of a space-slice at constant global time is zero; in other words, the Universe is flat, and the density parameter $\Omega_{total} = 1.0$.

Following the inflationary period, the Universe continues to expand, but at a much slower rate. The inflationary model helps to explain the large-scale structure of the Universe, because at this early stage, during the inflationary period, the Universe is expanding extremely rapidly, so this expansion essentially prevents the particle pairs from annihilating. In fact, the particle pairs are virtual, but the expansion of the Universe causes the particles to move apart from each other very rapidly. As a result, they cannot annihilate, and they are transformed into real particles. The Universe is hot and energetic enough at this time, and so this is where equation (2.2) comes into play. The energy here can give the particles their mass, and hence the Universe if now filled with real particles. This is the source of quantum fluctuations, which grew very rapidly to become overdense regions that provided the seeds for gravitational collapse. These collapsing overdensities were composed mainly of dark matter, and they provided the mechanisms necessary for visible (or normal) matter to condense to begin the process of galaxy formation. The idea is that these initial perturbations grew to become the galaxies and clusters of galaxies that we see in the current Universe. The inflationary model of the Universe was first developed in the late 1970s and early 1980s by physicists Alan Guth and Andrei Linde.

There is a slight problem with this somewhat over simplistic description of the early Universe. In this model, for every particle that is created, there should be an antiparticle. In that case, where is all the antimatter, and why do we live in a Universe dominated by matter? There are some symmetry violations that have been observed in particle physics experiments, and as a result there is slightly more matter created than antimatter in the early Universe. This idea will be described in more detail in chapter 4.

The overdensities created by quantum fluctuations in the inflationary model are the seeds for structures (or galaxies and clusters) to grow through gravitational collapse. Observations provide abundant evidence that the structure in the Universe formed hierarchically. In other words, the first structures (or galaxies) in the Universe were small. In an overdense region, the gravitational attraction between these small galaxies would result in them eventually merging. As a result, small structures evolve into larger galaxies over time.

The first dark matter halos begin to form as a result of quantum fluctuations and, in the present-day Universe, visible galaxies live at the centers of these massive halos. Dark matter dominates the total matter density of the Universe, and, as it does not interact with radiation, it is the first matter to undergo collapse due to gravity. Early halos of dark matter will grow through two processes. The first of these is smooth accretion of additional material. The second growth mechanism is mergers with other dark matter halos. In the standard cosmological models the merger rate of distinct dark matter halos is robustly predicted.

Figure 2.3. An image of the Antennae Galaxies, the closest example of a merger between two galaxies in the Universe. Image credit: NASA/STScI Hubble Space Telescope Legacy Team.

We expect that galaxies formed inside larger dark matter halos. The early galaxies begin to form as the dark matter halos draw baryonic matter gravitationally into the halos, allowing galaxies to form earlier than would otherwise be possible. These galaxies grow through mass accretion as additional material is drawn into the dark matter halo. Galaxy mergers allow the growth of more massive galaxies as the Universe continues to age. These merger events are key in altering galaxy morphologies by growing bulges in spiral galaxies, transforming spiral galaxies into elliptical galaxies and inducing star formation. In fact, from observations from the Hubble Space Telescope, we know that mergers between galaxies are common in the Universe (see for example, figure 2.3), and they were even more common when the Universe was young and galaxies were closer together.

Comparisons between predictions from models and simulations of a cold dark matter Universe and observations show remarkable agreement on the largest scales. Computer simulations of large volumes of the Universe in this cosmological model reproduce the clustering and distribution of galaxies on the largest scales, and they look incredibly homogeneous and isotropic. One of the most famous simulations of this kind is the Millennium Simulation[1], which was led by a group of astrophysicists at the *Max-Planck-Institut für Astrophysik* in Garching, Germany. Part of this simulation is shown in figure 3.1 at the beginning of chapter 3.

Despite the successes of the cold dark matter cosmology at describing the large-scale structure of the Universe, the model remains far from perfect. Observations show significant differences on small-scales from the theoretically predicted structures.

---

[1] http://www.mpa-garching.mpg.de/galform/virgo/millennium/

A detail of significant difference comes from how dark matter is distributed in galaxies, or within a dark matter halo. The density of dark matter will decline as we move out from the center of a halo. The exact formula, which describes how the dark matter density falls off as a function of radius (the density profile), is predicted by theoretical simulations of large-scale structure in the Universe (such as the Millennium Simulation). These simulations of structure growth have shown that galaxy-sized dark matter halos all have cuspy central densities, and computer simulations have shown that the dark matter density profiles, $\rho(r)$, within a halo all follow the same form (regardless of their mass or size), which is given by

$$\rho(r) = \frac{\rho_0}{\dfrac{r}{R_s}\left(1 + \dfrac{r}{R_s}\right)^2} \tag{2.4}$$

where r is the distance from the galactic center, and ρ_0 and R_s (the scale radius) are parameters which vary from halo to halo. Equation (2.4) is known as the Navarro, Frenk, and White (or NFW) dark matter density profile. From equation (2.4), it can be seen that the density is maximum in the very central regions of a galaxy, and it falls off very rapidly as you move out of the center (this is called a density cusp). However, observations of real galaxies, particularly the small, dwarf galaxies, show that the densities remain constant in the central regions, and the density only declines after a significant distance has been traversed (figure 2.4 demonstrates these differences), which is best given by the following, pseudo-isothermal density profile

$$\rho(r) = \frac{\rho_0 r_0^3}{(r + r_0)\left(r^2 + r_0^2\right)} \tag{2.5}$$

where r_0 is the core radius and ρ_0 is the central density. This discrepancy between dark matter simulations and observations of galaxies is still something that is difficult for cold dark matter cosmology to predict.

A potential solution for this conflict could be that the observational results are affected by a significant amount of non-circular motion in the central regions of a galaxy. In the determination of the observed density distribution of dark matter in galaxies, it is assumed that orbits of visible material (stars or gas) are circular (see equation (2.1), which is based upon circular motions). However, this is not necessarily the case, and orbits could actually be elliptical in nature. This would result in non-circular motions in the central regions of galaxies. There are other reasons why orbits could be non-circular in central regions. Some galaxies have bars, which would also induce non-circular motions. Most galaxies will almost certainly live in triaxial halos, in which the halo coordinates are given by

$$\frac{x^2}{a^2} + \frac{y^2}{b^2} + \frac{z^2}{c^2} = 1 \tag{2.6}$$

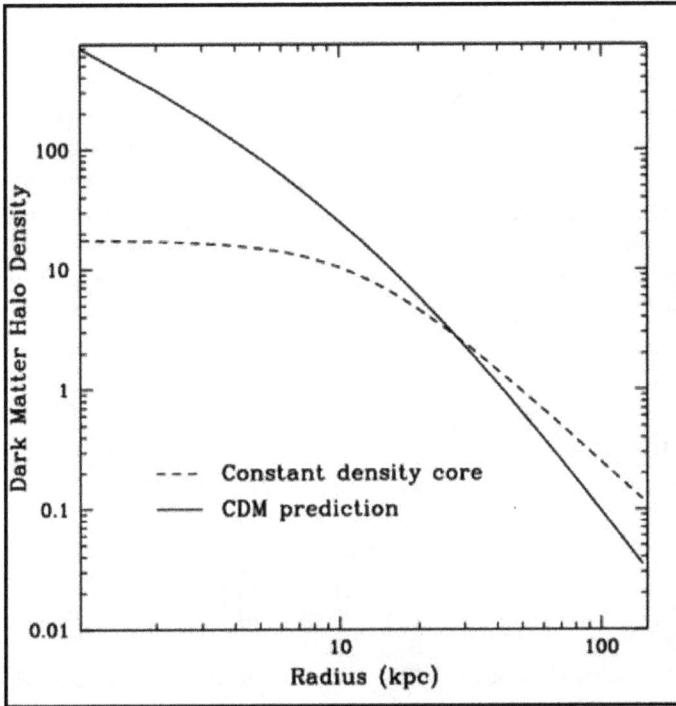

Figure 2.4. The difference between a density profile with a central cusp (the solid line or cold dark matter, CDM, prediction, also known as the NFW profile) and a density profile that remains constant in the central regions (the dashed line or constant density core).

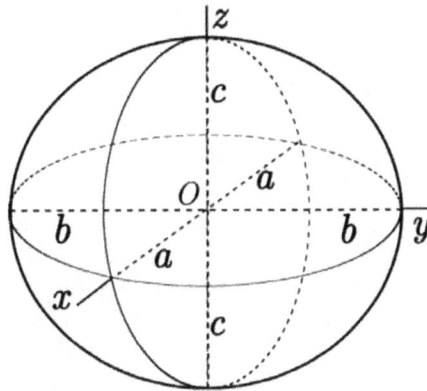

Figure 2.5. A triaxial halo or ellipsoid.

where x, y, z are Cartesian coordinates and a, b, c are the semi-principal axes in the x-, y-, and z-directions, respectively. An example of a triaxial halo (or ellipsoid) is shown in figure 2.5.

All of these possibilities result in non-circular motion, which means that a star will change its speed as it moves along its orbit (unlike circular motion, where its speed remains constant). The velocity of a star in its orbit is described by Kepler's second law, which states that an imaginary line joining that star to the gravitational center of the Galaxy sweeps out an equal area of space in equal amounts of time (Kepler's original laws were initially applied to planets, but Newton showed they could be applied to orbits in any gravitationally bound system). For elliptical orbits, Kepler's second law implies that stars move fastest when they are at their closest approach to the center of the galaxy. As the distance of a star from the galactic center changes constantly if orbits are elliptical, the stellar velocities are also constantly changing.

The density we calculate depends upon the speed of the objects we are using to trace the mass. This is fine for circular motions, but if non-circular motions are present, and speeds are changing depending on where a star is in its orbit, then our density calculations would be incorrect. Instead of using equation (2.1) to determine the central mass and therefore density profile, we need to know the exact shape of the elliptical orbits, and we need to use an equation that relates orbital velocities to the shapes of the elliptical orbits rather than assuming circular motions (see equation (2.7)), such that

$$v_{\text{rot}} = \sqrt{GM\left(\frac{2}{r} - \frac{1}{a}\right)} \qquad (2.7)$$

where M is the enclosed mass at radius r and a is the semi-major axis of the elliptical orbit. To determine if orbits are elliptical, and the eccentricity of the orbits, we need instrumentation that can map the velocity fields of galaxies, and this can be done with integral field unit (IFU) spectroscopy. These types of instruments, however, typically can only do this for the very smallest (i.e., dwarf) galaxies or for the very central regions of large galaxies.

These non-circular (or elliptical) motions may result in observations of lower than expected velocities in the central regions of galaxies. Despite these differences in how orbital shape affects velocities of stars, several studies (particularly for dwarf galaxies) have shown that, even when taking into account these non-circular motions, it is difficult to reproduce the dark halo densities expected by cold dark matter cosmology.

Dwarf galaxies are dominated by their dark matter content at all locations. In other words, the visible material is such a small amount of matter in these galaxies that estimating the distribution of the visible mass incorrectly does not strongly affect our estimates of the dark matter density distributions. Their velocity fields can also be mapped with IFU spectrographs. Dwarf galaxies are, therefore, great laboratories with which to study dark matter halos. Most studies of dwarf galaxies conclude that dark matter models are in clear conflict with the density distribution of dwarf galaxies. One of the exceptions is the Triangulum Galaxy, M33, a small satellite galaxy in orbit around the Andromeda Galaxy. The Triangulum Galaxy may be the only small galaxy which has a cuspy central density (and therefore appears to be consistent with the cold dark matter paradigm) that we know of (see figure 2.6). It is a very late-type galaxy, and we therefore might expect it to have a constant density core, just like the dwarf galaxies described above. Nevertheless, two

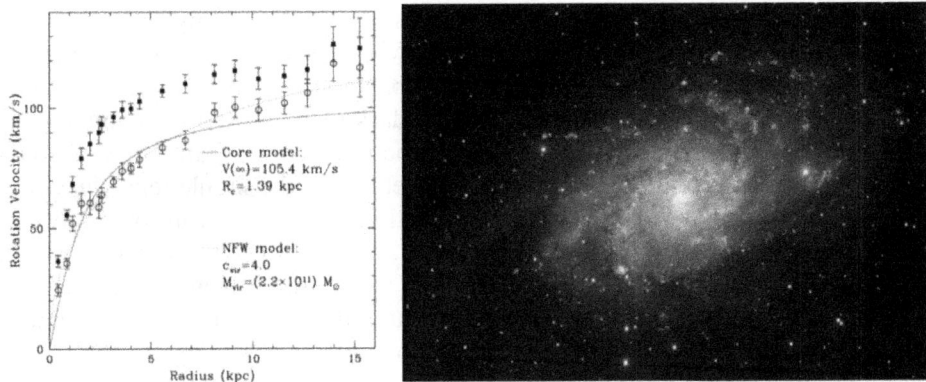

Figure 2.6. The rotation curve of the Triangulum Galaxy (left) modeled using a pseudo-isothermal model (solid blue line) and the dark matter simulation model (or NFW model, dotted red line). The squares are the total rotation velocities of the neutral hydrogen gas in the Triangulum Galaxy. The circles are the dark matter contribution to the rotation velocities after subtraction of the stellar and gas mass. The rotation curve is taken from Seigar (2011, ISRN: Astronomy and Astrophysics, Vol. 2011, Article ID 275697, 8 pages). Image of the Triangulum Galaxy (right). Image credits (right): European Southern Observatory's VLT Survey Telescope.

studies of M33 from the last decade both show that it seems to conform to the expectations of cold dark matter cosmology.

Despite the work on the Triangulum Galaxy pointing towards a dark matter distribution that is consistent with cold dark matter cosmology, there are many other studies (already mentioned above) that clearly indicate that there is a problem. Several possible solutions to resolving this conflict have been suggested. One possibility is that these observations are pointing to a real problem with cold dark matter cosmology, perhaps indicating that dark matter is not cold, but rather warm (see chapter 3 for a clearer explanation of cold versus hot dark matter), or possibly even a form of self-interacting dark matter, in which case it is easier to produce constant density cores at the centers of dark matter halos. In the last few years, however, it has been shown that these scenarios are unlikely. This is because of discrepancies in the sizes of the cores created in the warm dark matter models, which are much smaller than the observed sizes. The conclusion is that there is no motivation to prefer warm dark matter halos over cold dark matter halos, at least from an astrophysical perspective.

Another possible explanation for the prevalence of constant density cores in these dwarf galaxies is because they formed late in the history of the Universe, unlike large elliptical and spiral galaxies, which formed at earlier times and therefore conform to the standard expectations cold dark matter cosmology. This is because the central mass densities of galaxies tend to reflect the density of the Universe at their formation time. Galaxies that formed when the Universe was younger will have higher central densities simply because the Universe was smaller, and therefore the overall density of the Universe was higher. Nevertheless, while this explanation may be appropriate for the lower central densities typically found in dwarf galaxies, it does not explain the shape of the inner density profiles, as the dark matter simulations presented for dwarf galaxies still result in central cuspy densities for all mass scales, albeit with a lower central density when compared to massive galaxies.

The last possible explanation for this discrepancy between theory and observations is that the simulations are based on pure dark matter and do not contain any physics relating to normal, visible matter, or the interaction (if any) between dark and visible matter. If one were to include these effects in the simulations, the visible matter may interact (at least gravitationally) with the dark matter in some way that might resolve this cusp/core problem. However, the effect of visible matter on the dark matter may act to make the problem even worse. As early as the mid-1980s it was shown that as visible matter cools and falls into the center of a dark matter halo to form a visible galaxy, the dark matter halo contracts. This makes the central density profile even more cuspy and also sets up other problems (for instance, in this scenario, galaxies of a given mass—or size—rotate faster in these simulations than our observations suggest). More recently, however, mechanisms for preventing this contraction have been suggested, and they may appear to work. These mechanisms are referred to as feedback, and there are two types of feedback.

The first type of feedback is feedback from star formation. Observations have made it clear that in many galaxies, stars form at a much lower rate than expected. A second observational puzzle is the presence of high velocity outflows of cold gas in galaxies that form stars at the highest rates. These outflows may be a result of winds from massive stars or supernova explosions. In some cases, the rate at which gas is outflowing is similar to the rates at which stars are forming. Two things can now happen. First of all, these observations suggest that, at least in some systems, the star formation provides a feedback mechanism that removes the neighboring gas that would otherwise be available for further star formation. This limits the size of the visible galaxy for a given halo mass. Secondly, this also means that the efficiency at which material (i.e., gas) falls to the center of a halo is decreased due to these winds or outflows. This changes the interaction between the normal matter and the dark matter, and as a result the above halo contraction no longer occurs.

The second form of feedback is called AGN feedback. Almost every galaxy in the Universe harbors a supermassive black hole in its nucleus. These black holes have masses of between 100 000 and a few billion times the mass of the Sun. At some point in the life of a galaxy, these supermassive black holes would have been actively swallowing (or accreting) material. Material falling into a black hole first forms a hot accretion disk around it. For supermassive black holes, these disks become so hot that they can outshine their host galaxies. This is an active galactic nucleus or AGN. Some of these AGN have jets of material that blast material away from the central regions at speeds close to the speed of light (typically the minimum velocities observed are about $0.9c$). These jets are another feedback mechanism that can both expel gas from a halo (and thus inhibit further star formation) and a mechanism that can stop halo contraction.

So, while theoretical results from the mid-1980s suggested that the interaction between visible and dark matter would lead to an even cuspier central profile, our understanding of feedback physics (which is supported by observations) has shown that this contraction probably does not take place. This alone does not resolve the cusp/core problem, but it also does not make it any worse than it has to be.

If feedback from star formation can affect the early stages of halo formation, steep cuspy density profiles may be transformed into the observed flat cores. In one particular set of simulations that included star formation feedback, strong outflows from supernova explosions remove gas, which inhibits the formation of central visible bulges in galaxies and decreases the dark-matter density in the central regions. Galaxies that are bulgeless and have shallow central dark-matter profiles arise from these simulations. These are the analogues of the late-type and dwarf galaxies that are observed to have constant density core profiles in their central regions. These simulations seem to provide a working solution to the cusp/core problem. From this work, it appears that supernova feedback provides a mechanism for transforming cuspy density profiles into shallower central density profiles. However, even in the case of these simulations, all of the observational studies that have dealt with the cusp/core problem have focused on the properties of late-type dwarf and low-surface brightness galaxies. Galaxies with dominant bulges have yet to be explored in detail. In the few cases where such galaxies have been explored, a cuspy profile seems to result from the observational data. Also, in the case of elliptical galaxies (which could be thought of as pure-bulge galaxies) and clusters of galaxies the cupsy cold dark matter model provides a remarkable fit to the density profile. This highlights the need to extend these studies to include galaxies with significant bulges. Such studies are difficult, because the central regions of these galaxies are dominated by visible matter (unlike dwarf galaxies which are dark matter dominated at all locations). Determining the central visible mass depends on several factors such as how accurate the light distribution can be converted into a mass distribution, which is definitely not simple as it depends on the types of stars that make up the light that we see. There are theoretical models that can help us calculate this transformation from light to stellar mass, but they are all approximations. In a region where the dark matter content is small, even the smallest inaccuracies or approximations will lead to very misleading results.

So, to bring this chapter to a conclusion, cold dark matter cosmology reproduces the large-scale structure of the Universe extremely accurately. However, on galaxy-sized scales there are several known issues with the theory. Here, we have highlighted one particular problem, namely the cusp/core problem, which highlights the fact that cold dark matter simulations of structure formation predict central cusps in the density profiles of dark matter halos. However, observations, particularly of dwarf and late-type galaxies, suggest that the central densities are flat over the inner 3000 light-years (about 1 kilo-parsec). This is a serious issue with cold dark matter cosmology. A solution involving supernova feedback may be the answer, but this model needs to be applied to large spiral and elliptical galaxies to see how widely applicable it is.

Suggested further reading

Albrecht A and Steinhardt P 1983 *Phys. Lett.* B **131** 45–8

Blumenthal G R, Faber S M, Flores R and Primack J R 1986 *Astrophys. J.* **301** 27–34

Bullock J S *et al* 2001 *Mon. Not. R. Astron. Soc.* **321** 559–75

Burkert A 2000 *Astrophys. J.* **534** L1436

Governato F *et al* 2010 *Nature* **463** 203–6

Guth A H and Tye S-H H 1980 *Phys. Rev. Lett.* **44** 631–35

Kuzio de Naray R and Kaufmann T 2011 *Mon. Not. R. Astron. Soc.* **414** 3617–26

LeMaitre G 1931 *Mon. Not. R. Astron. Soc.* **91** 483–90

Linde A D 1982 *Phys. Lett.* B **116** 335–39

Navarro J F, Frenk C S and White S D M 1996 *Astrophys. J.* **462** 563–75

Navarro J F, Frenk C S and White S D M 1997 *Astrophys. J.* **390** 493–508

Navarro J F *et al* 2004 *Mon. Not. R. Astron. Soc.* **349** 1039–51

Seigar M S, Barth A J and Bullock J S 2008 *Mon. Not. R. Astron. Soc.* **389** 1911–23

Seigar M S 2011 *ISRN Astron. Astrophys.* **2011** 725697

Seigar M S and Berrier J 2011 Galaxy rotation curves in the context of ΛCDM cosmology *Advances in Modern Cosmology* pp 77–102

Chapter 3

Cold dark matter, hot dark matter, and their alternatives

In this chapter, we will learn that there are two broad types of dark matter, hot dark matter and cold dark matter (and maybe a third in the form of warm dark matter). We will learn why dark matter is dominated by cold dark matter, and why alternatives to dark matter (such as modified theories of gravity) cannot explain some of the observed features of galaxies and large-scale structures that we see in the Universe.

What is cold dark matter? What is hot dark matter? The distinguishing characteristic in determining the 'temperature' of matter is how fast it moves. In thermodynamics, the kinetic energy of a gas is related directly to its temperature as follows

$$E_k = \frac{3}{2}kT \tag{3.1}$$

where E_k is the kinetic energy, T is the temperature in Kelvin, and $k = 1.38 \times 10^{-23} \text{ m}^2 \text{ kg s}^{-2} \text{ K}^{-1}$ is Boltzmann's constant.

Kinetic energy is the energy of motion. The faster something moves, the more kinetic energy it has. In fact, the kinetic energy is related to the square of the speed at which something is moving, as follows

$$E_k = \frac{1}{2}mv^2 \tag{3.2}$$

where m is the mass of a particle and v is its velocity. So, if something is fast moving, it is 'hot', if something is slow moving, it is 'cold', and something that is moving at moderate speeds could be considered 'warm'. To demonstrate this mathematically, we can manipulate equations (3.1) and (3.2) and show that temperature and velocity are, in fact related to each other as

$$v_{rms} = \sqrt{\frac{3kT}{m}} \tag{3.3}$$

doi:10.1088/978-1-6817-4118-5ch3

where v_{rms} is the root mean square of the velocity, a way of measuring the 'average' speed of particles in a gas.

The fact that particle speeds and the temperature of a gas are related to each other through equation (3.3) is a basic feature of matter, but it does not take into consideration a fundamental, universal speed limit. In 1905, with the publication of his theory of special relativity, Einstein showed that the speed of light (186 282 miles per second or 299 792 458 meters per second) is a fundamental limit in the Universe. Nothing can travel faster than light. Does this change our definition of 'hot' versus 'cold' matter? The simple answer to this question is, yes! However, now all it means is that any material/gas with speeds that are a significant fraction of the speed of light is 'hot' or relativistic, whereas slow moving material (typically moving at less than ten percent of the speed of light) is considered 'cold'. So now we have put a boundary on the speeds, and we have better defined what we mean by fast and slow or hot and cold.

This brings us back to the original question. At least now, we can say that cold dark matter moves at low speeds (which allows us to perform simulations of the type seen in figure 3.1), typically less than ten percent of the speed of light, and hot dark matter moves at speeds typically greater than ten percent of the speed of light. Let us take a look at hot dark matter first. In fact, we will take a look at a particular class of particles called neutrinos, essentially because they were first proposed by Wolfgang Pauli in 1930. They were first detected experimentally in 1956 by Clyde Cowan and Frederick Reines, a breakthrough for which Reines won the Nobel Prize in Physics in 1995 (the physics building at the University of California, Irvine is named in his honor).

A neutrino is an electrically neutral weakly interacting elementary subatomic particle. During experiments in the 1910s and 1920s, certain types of particles, called beta particles, were first observed as a result of radioactive decay. In fact, a beta particle is simply an electron. The strange thing about the beta particles that were being emitted, is that they had a range of velocities, which means that their kinetic energies were very different. However, we know that energy has to be conserved, which means that another, unseen particle must be carrying away the rest of the energy. This led Pauli to postulate the existence of the neutrino. In this radioactive beta decay a neutron (in an atomic nucleus) decays into a proton (which stays in the nucleus), an electron and an anti-neutrino (which are emitted from the nucleus) as follows:

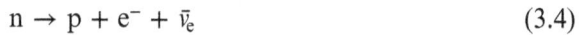

$$n \rightarrow p + e^- + \bar{v}_e \tag{3.4}$$

In the above equation, n is a neutron, p is a proton, e^- is an electron, and $\bar{v}_e$ is an anti-neutrino. We now know that neutrinos can be created in several ways, including in certain types of radioactive decay, in nuclear reactions such as those that take place in the Sun, in nuclear reactors, and when cosmic rays hit atoms. In the Sun, the proton–proton chain reaction begins with the fusion of two protons into a helium-2 nucleus, which immediately undergoes beta decay to a deuterium nucleus (which consists of a proton and a neutron), a positron and an electron neutrino. The process conserves the lepton number by producing a lepton (the neutrino) and an antilepton (the positron), as shown in figure 3.2.

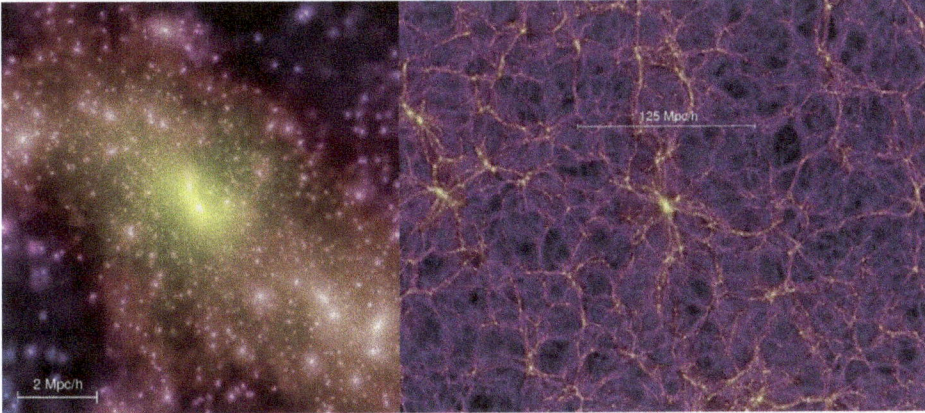

Figure 3.1. A cold dark matter simulation of structure in the Universe. *Images courtesy of The Millennium Simulation Project, Max-Planck-Institut für Astrophysik* (Springel *et al* 2005 *Nature* **435** 7042 629–36).

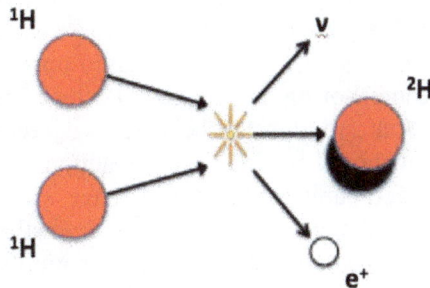

Figure 3.2. The first step of the proton–proton chain, in which two hydrogen nuclei (two protons, denoted ^{1}H in the figure) fuse to form a deuterium nucleus (^{2}H in the figure). The deuterium nucleus consists of a proton (red circle) and a neutron (black circle). One of the protons has undergone beta decay and has transformed into a proton. In order to conserve charge and lepton number, the beta decay of one of the protons also results in a positron (e^+) and a neutrino (v).

We also now know that there are three types, or 'flavors', of neutrinos: electron neutrinos, muon neutrinos, and tau neutrinos (the neutrino in equation (3.4) is actually an electron anti-neutrino and the neutrino in figure 3.2 is an electron neutrino). Each type is associated with an antiparticle, called an 'anti-neutrino', which also has neutral electric charge and half-integer spin. It is even possible for neutrinos to change flavor or 'oscillate', and this experimental result suggests that neutrinos do have mass, even if it is very small (see table 3.1 for the upper mass limit of neutrinos). This is an important point, because neutrinos interact 'weakly' with other matter, which is a key feature of dark matter. Dark matter is weakly interacting, meaning that it rarely interacts with other matter, except in the gravitational sense. This is also true of neutrinos (although their masses are very small compared to cold dark matter candidates). Nevertheless, they do contribute something to the mass budget of the Universe, and they have therefore been proposed as a form of hot dark matter.

Table 3.1. Flavors of neutrinos and their mass upper limits.

Neutrino type	Symbol	Mass limit (kg)	Mass limit (eV)
Electron neutrino	ν_e	3.9×10^{-36} kg	2.2 eV
Muon neutrino	ν_μ	3.0×10^{-31} kg	170 keV
Tau neutrino	ν_τ	2.8×10^{-29} kg	16 MeV

While neutrinos have a small mass, there are an incredibly large number of them in the Universe. For example, every second about 1.8×10^{38} neutrinos are emitted by the Sun. The Sun is just one of about 200 billion stars in the galaxy, which means that the galaxy produces about 3.6×10^{49} neutrinos every second (assuming the Sun is a fairly average star). The Milky Way galaxy has been around for about 10 billion years, which means that over its entire lifetime, it has produced about 1.1×10^{67} neutrinos. The Milky Way is just one of about 100 billion galaxies in the Universe. If the Milky Way is a typical galaxy, this means that there should be about 1.1×10^{78} neutrinos flying around the cosmos. This is just based upon neutrinos produced by nuclear reactions in the centers of stars. It seems like a lot of neutrinos. However, the nuclear reactions that occurred during the first two seconds of the Big Bang produced a lot more neutrinos than this. These reactions in the Big Bang created enough neutrinos to fill each cubic centimeter of space with at least 150 neutrinos. Given the current volume of the Universe, this is equivalent to a total of about 4.6×10^{86} neutrinos (that is 400 million more than the number of neutrinos created in stars). If these were all the lightest form of neutrino (the electron neutrino), the mass limits above give a maximum possible mass of 1.8×10^{51} kg from neutrinos in the Universe. Our current estimate of the amount of visible mass in the Universe stands at about 10^{53} kg. If you include dark matter in the Universe, the total mass in the Universe increases to about 10^{54} kg. So neutrinos account for less than 0.18% of the mass budget of the Universe. In this calculation, we assumed that all neutrinos in the Universe are electron neutrinos, i.e., the lightest 'flavor' of neutrino. This is realistic, to some degree, since most particles tend to decay to the lightest stable particle. However, there should be some muon neutrinos that we should take into account. The numbers of tau neutrinos are probably negligible and can be ignored. Still, for neutrinos, we are talking about a particle that accounts for a few tenths of a percent of the mass in the Universe. It turns out that, while this seems small, it is not insignificant.

Let us turn our sights now to cold dark matter. Most of the mass (almost 90% of it) in the Universe is thought to be in the form of cold dark matter. Why do we believe this to be the case? Why can it not be neutrinos (a form of hot dark matter)?

Hot dark matter was popular for a time in the early 1980s. However, it suffers from a severe problem. This is a problem referred to as free-streaming length. As we have already discussed, dark matter is hot or cold (or warm) depending on the speed at which it was moving, and this is related to how far it moved due to random motions in the early Universe before it could be slowed down due to the expansion of the Universe. This is an important distance called the 'free-streaming length'. Primordial density fluctuations smaller than this free-streaming length get washed out as particles move from overdense to underdense regions, while fluctuations greater than this distance are unaffected. The free-streaming length effectively sets a minimum scale for structure formation. The problem with hot dark matter (including neutrinos) is that its free-streaming length is much larger than the size of a galaxy. If all the dark matter were hot, all galaxy-sized density fluctuations would get washed out due to free streaming. In hot dark matter models this means that the first structures that can form are huge supercluster-sized pancakes, which were then theorized to somehow fragment and break up into smaller clusters and further break up into galaxies. However, observations of deep fields clearly show that galaxies formed at very early times, with clusters and superclusters forming later as galaxies clumped together under their mutual gravitational attraction. Therefore, any model dominated by hot dark matter is seriously in conflict with observations.

Cold dark matter on the other hand has a negligible thermal velocity when compared to the expansion of the Universe (or the Hubble flow). Its non-gravitational interactions are much weaker than the weak interactions. Cold dark matter particles have free-streaming lengths that are much smaller than the typical size of an ancestor of a galaxy-sized perturbation. This is currently the area of greatest interest for any research into dark matter, and most particle candidates become non-relativistic at very early times and are hence classified as cold. With cold dark matter, galaxies can form at early times, and they can clump into clusters and superclusters over time through their mutual gravitational attraction. This is what we observe in deep field surveys, which is why cold dark matter is so popular.

As mentioned in chapter 2, there are some problems with cold dark matter, particularly when it comes to describing the central mass densities of dark matter in galaxies. Cold dark matter predicts central cusps, whereas observational results prefer central cores, an issue known as the cusp/core problem. Warm dark matter has been proposed as a possible solution to this discrepancy. Warm dark matter refers to particles with a free-streaming length comparable to the size of a region, which subsequently grew to become a dwarf galaxy. The predictions of warm dark matter are similar to cold dark matter on large scales (including the cosmic microwave background, galaxy clustering, and large galaxies), but with fewer small-scale density perturbations. This reduces the number of dwarf galaxies (cold dark matter overproduced dwarf galaxies when compared to observations) and it also leads to lower densities in the central regions of galaxies, and indeed produces a constant density core. However, warm dark matter models predict that the core radii should decrease with halo mass, whereas observations find that the core radii increase with halo mass. Furthermore, dark matter halo densities can vary by a factor of about 30 from galaxy to galaxy, and this is independent of the maximum rotation curve of the galaxy. Therefore, these cores do not provide evidence to support warm dark matter over cold dark matter. Another challenge for the warm dark matter model is that there are no well-motivated particle physics candidates with the required mass range, which is about 300–3000 eV. There is a postulated candidate for warm dark matter, which is the sterile neutrino. This is a heavier, slower form of neutrino, which does not even interact through the weak nuclear force (unlike regular neutrinos). Also, the gravitino can have properties similar to warm dark matter, although strictly speaking, it is a cold dark matter particle candidate.

Another model proposed in the early 1980s is that of mixed dark matter, and this is now considered obsolete. The mixed dark matter model was specifically chosen to have a mass ratio of 80% cold dark matter and 20% hot dark matter (neutrinos). We know that hot dark matter coexists with cold dark matter in any case, but there was a very specific reason for choosing this ratio of hot to cold dark matter. During the early 1990s it became steadily clear that a Universe with critical density of cold dark matter did not fit the observations of the cosmic microwave background (particularly from NASA's Cosmic Background Explorer, COBE). There was also a problem with the observed large-scale structure and clustering. These problems could be reconciled in two ways. The first was mixed dark matter with an 80/20 mixture ratio, the second was the introduction of a cosmological constant, essentially a repulsive force that pushes out against gravity (this latter idea is the so-called ΛCDM model). With the discovery of the accelerating Universe from Type 1a supernovae data in the late 1990s, this discrepancy was essentially resolved. The acceleration of the Universe can be explained by the introduction of a cosmological constant, Λ, on top of the cold dark matter (CDM) model, so our preferred cosmological model is now the concordance ΛCDM model.

There are some alternative theories to the 'missing mass problem' that do not include dark matter. The most commonly used models that are used as an alternative to dark matter are theories of modified gravity. They modify the laws of gravity established by Newton and Einstein.

The earliest modified gravity model to emerge was Mordehai Milgrom's modified Newtonian dynamics (MOND) in 1983, which adjusts Newton's laws to create a stronger gravitational field when gravitational acceleration levels become tiny (such as near the rim of a galaxy). It had some success explaining galactic-scale features, such as rotational velocity curves of elliptical galaxies, and dwarf elliptical galaxies, but did not successfully explain galaxy cluster gravitational lensing. However, MOND was not relativistic, since it was just a straight adjustment of the older Newtonian account of gravitation, not of the newer account in Einstein's general relativity. Soon after 1983, attempts were made to bring MOND into conformity with general relativity; this is an ongoing process, and many competing hypotheses have emerged based around the original MOND model.

Most (if not all) MOND models have essentially been ruled out since the discovery of the Bullet Cluster (see figure 3.3). The Bullet Cluster consists of two colliding clusters of galaxies.

The major components of the Bullet Cluster (stars, gas, and dark matter) behave differently during the collision and this allows them to be studied separately. The stars of the galaxies, which are observable in visible light, were not greatly affected by the collision, and most passed right through. They were slowed gravitationally but not otherwise altered. The Chandra X-ray Observatory telescope highlights the hot gas

Figure 3.3. The Bullet Cluster: Hubble Space Telescope image with overlays. The total projected mass distribution reconstructed from strong and weak gravitational lensing is shown in blue, while the x-ray emitting hot gas observed with Chandra is shown in red. Image courtesy of NASA's Hubble Space Telescope and Chandra Telescope Great Observatories.

in the two colliding cluster components. This actually represents most of the ordinary (or visible) matter in the Bullet Cluster. The hot gases interact electro-magnetically, and this causes the gases of both clusters to slow much more than the stars. The third component, the dark matter, was detected indirectly by the gravitational lensing of background objects. Gravitational lensing is a prediction of General Relativity. Light from background objects is bent around an intervening mass (in this case the dark matter in the Bullet Cluster) on its way to Earth (see figure 3.4 for a diagram of how this works). This results in distorted or multiple images of the background objects. In theories without dark matter, such as MOND, the lensing would be expected to follow the ordinary matter or the hot x-ray gas. However, the lensing is strongest in two separated regions near (possibly coincident with) the visible galaxies. This provides support for the idea that most of the mass in the cluster pair is in the form of two regions of dark matter, which bypassed the gas regions during the collision. This observational result agrees with predictions of cold dark matter, and is probably the best experimental data that we have for confirmation of the cold dark matter model.

The Bullet Cluster provides the best current evidence for the nature of dark matter, but, more importantly, it provides evidence against MOND, at least as applied to large galactic clusters. The spatial offset of the center of the total mass from the center of the visible mass peaks cannot be explained with an alteration of the gravitational force law alone, i.e., by changing Newton's or Einstein's laws of gravity.

The cluster is undergoing a high-velocity merger at about 4500 km s^{-1}, and this measurement has been made from the spatial distribution of the x-ray emitting gas

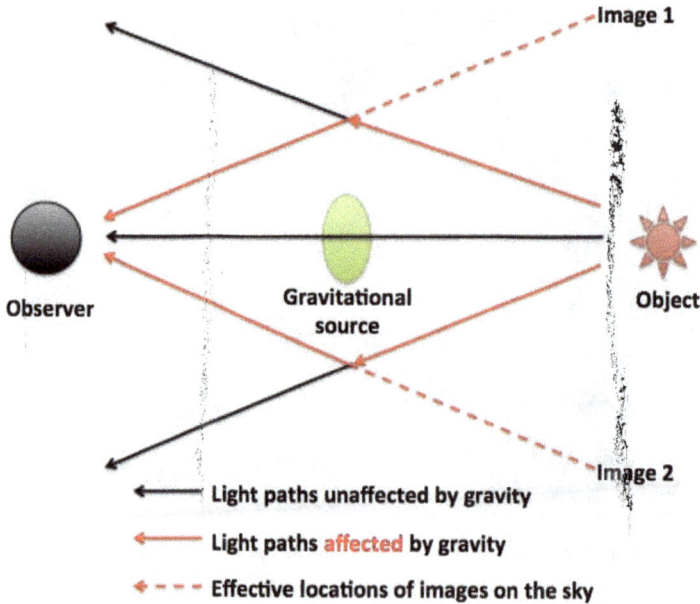

Figure 3.4. A schematic view of how gravitational lensing works. If there is not a foreground gravitational source, the black lines show how the light travels from the distant object and reaches Earth. If a gravitational source lies between Earth and the distant object, the red lines show how the light path bends and travels around the gravitational source. As a result, the distant object is amplified and another path of light reaches us. If we trace the red lines arriving at Earth back in a straight line (red-dashed lines) to the same distance as the object, this tells us where the images of the object will appear in the sky.

and how it lags behind the two clusters. The dark matter clumps (which are located via gravitational lensing) are coincident with the visible galaxies, and lie ahead of the hot x-ray gas. This has been used to put constraints on the type of dark matter in the Bullet Cluster, at least in terms of its interaction cross-section. The substructure velocities in the Bullet Cluster are not particularly high, as is expected for cold dark matter cosmology, and this scenario favors concordant ΛCDM cosmology. While the Bullet Cluster provides evidence for cold dark matter on cluster scales, the observed ratio of dark matter to visible matter in the Bullet Cluster is lower than predicted by the cold dark matter model. This may indicate that ΛCDM cosmology is insufficient to describe the mass discrepancy on galaxy scales, or that its predictions about the shape of the Universe are incorrect.

In conclusion, while alternatives have been proposed to dark matter, the only model that appears to be able to create the kinds of structures that we see in the Universe (at least in terms of objects of cluster size or larger) is that of cold dark matter. To describe large-scale structure on scales comparable to that of the observable Universe, we need to include a cosmological constant, but with the discovery of the accelerating Universe in the late 1990s, this is now the preferred model, one that is referred to as ΛCDM cosmology. The cosmological constant is often referred to as dark energy, and it is a problem that is distinct from that of dark matter. The two are not related, but it appears that they are both necessary. This will be a topic that will be covered in chapter 7 of this book.

Suggested further reading

Bekenstein J D 2004 *Phys. Rev.* D **70** 083509

Blumenthal G R, Faber S M, Primack J R and Rees M J 1984 *Nature* **311** 517–25

Clowe D *et al* 2006 *Astrophys. J.* **648** L109–13

Efstathiou G and Silk J 1983 *Fundamentals of Cosmic Physics* (London: Gordon and Breach) pp 91–138

Milgrom M 2015 *Can. J. Phys.* **93** 107–18

Springel V *et al* 2005 *Nature* **435** 629–36

Chapter 4

Types of dark matter

We will now look at the dark matter problem from a particle physics point of view. We will do this by introducing the Standard Model of particle physics and describing some of its problems and shortfalls. Extensions to the Standard Model will be introduced, in particular those models, which predict the existence of new particles that have the expected properties of dark matter. Due to the complex nature of this material, this chapter includes more mathematics, and as a result, some readers may find this chapter more advanced.

The Standard Model of particle physics is a spectacularly successful theory of elementary particles and their interactions. It describes three of the fundamental interactions, the electromagnetic interaction, the weak nuclear interaction, and the strong nuclear interaction (the fourth fundamental interaction is gravity, but it is not included in the Standard Model). These interactions mediate the dynamics between all of the known subatomic particles. Subatomic particles fall into four categories. Leptons consist of electrons, muons, and tau particles, and their neutrinos (neutrinos come in three 'flavors', the electron neutrino, the muon neutrino, and the tau neutrino), so there are a total of six leptons (shown in green in figure 4.1). Quarks consist of the up and down quark, the charm and strange quark, and the top and bottom quark, so there are a total of six quarks (shown in blue in figure 4.1). There are four gauge bosons, the photon, which mediates the electromagnetic interaction, the W and Z bosons, which mediate the weak nuclear interaction, and gluons, which mediate the strong nuclear force (shown in orange in figure 4.1). The fourth category of particle is the Higgs boson, which falls in a category all of its own (shown in purple in figure 4.1). The classification scheme, which is essentially the Standard Model of particle physics, is well summarized by the schematic shown in figure 4.1. In this scheme, leptons and quarks form a class of particles known as fermions. Fermions are distinct from bosons in the amount of spin (or angular momentum) that they can carry (fermions have half-integer spin, whereas bosons have integer spin). As a result of this difference in spin, fermions and bosons have very different

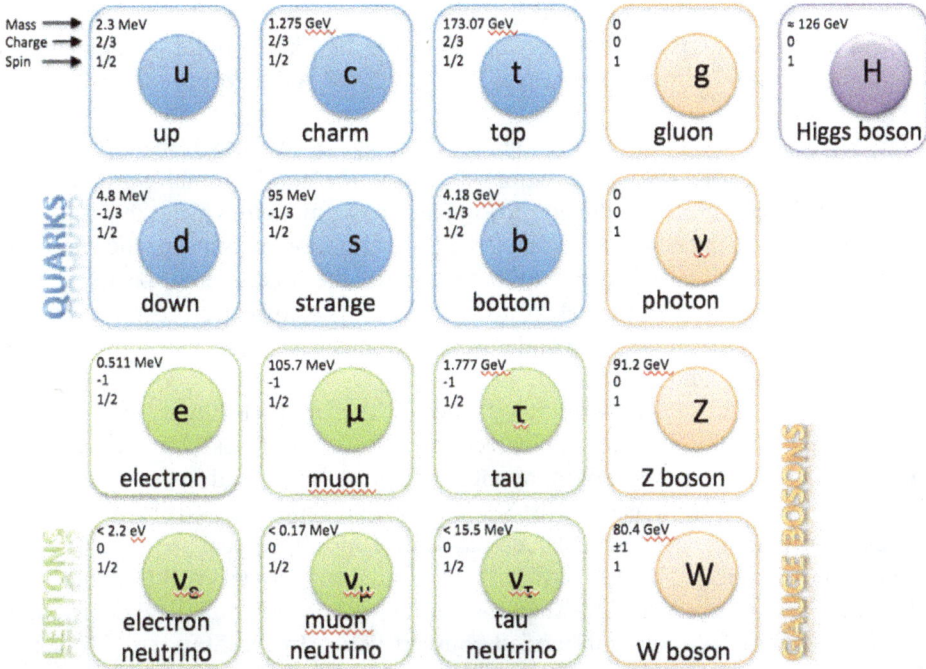

Figure 4.1. The Standard Model of particle physics. Quarks are shown in blue, leptons and neutrinos in green, gauge bosons (or force carriers) are shown in orange, and the Higgs boson is shown in purple.

properties, and this is very important, especially for some of the models that expand the Standard Model, in particular, the supersymmetric models.

However, the Standard Model of particle physics is not complete. One major problem of the Standard Model is that it does not include gravity (as described above), one of the four fundamental forces (or interactions). The model also fails to explain why gravity is so much weaker than the electromagnetic or nuclear forces. For example, a simple fridge magnet can counteract the gravitational attraction of a whole planet on a small object. Another example is that a falling object can take several seconds to reach the ground due to gravitational attraction, but the electromagnetic interaction between its and the ground's subatomic particles, will stop it almost immediately when it reaches the ground. As noted above, the Standard Model does not actually include gravity. The hypothetical gauge boson that mediates the gravitational force is called the graviton. If it exists, the graviton is expected to be massless (because the gravitational force appears to have unlimited range) and it must be a spin-2 boson.

The huge difference in the strength of fundamental forces is one aspect of the 'hierarchy problem'. It also refers to the wide range in mass for the elementary particles. In figure 4.1 shown above, the electron is about 200 times lighter than the muon and 3500 times lighter than the tau. Quarks also have wide ranges in mass: the top quark is 75 000 times heavier than the up quark. The Standard Model of particle physics does not explain why there is such a wide spectrum of masses among the building blocks of matter. Imagine what it would be like to build a house out of bricks with this kind of mass difference! The likelihood is that your house would fall down.

The hierarchy problem is also related to the Higgs boson mass. The equations of the Standard Model establish relations between the fundamental particles. For example, in the equations, the Higgs boson has a basic mass to which theorists add a correction for each particle that interacts with it, with more massive particles having larger corrections. The top quark is the heaviest of the fundamental particles, so it has the largest correction. In fact, it adds such a large correction to the predicted Higgs boson mass that theorists wonder how the measured Higgs boson mass can be as small as it was found to be at the LHC.

This seems to indicate that other yet undiscovered particles exist, particles that change the picture. In that case, the corrections to the Higgs mass from the top quark could be cancelled out by some other hypothetical particle and lead to the observed low Higgs boson mass. An extension to the Standard Model, known as supersymmetry just happens to predict the existence of such particles, hence its appeal.

One other major problem exists with the Standard Model, the neutrino mass problem. The masses of fermions (leptons and quarks, for example) are described in field theories by left- and right-handed coupled fields. In the Standard Model, all neutrinos are left-handed, so the Standard Model predicts that neutrinos are massless. The Super-Kamiokande experiment was able to demonstrate that neutrinos can change flavor (e.g., from electron neutrino to muon neutrino) and this result implies that the three flavors of neutrino are non-degenerate, which means that they should have a mass. In relativistic physics, only massless particles can travel at the speed of light (for example, the photon, and the hypothesized graviton). A particle that travels at light speed cannot change any of its characteristics. For example, a photon has the same characteristics/properties when it arrives at the Earth as it had when it left the Sun. This means that if a particle can change its characteristics (or in the case of neutrinos, change its flavor) it must be traveling at a speed slower than that of light, and therefore it has mass. The fact that neutrinos have mass provides the most direct and compelling evidence that the Standard Model of particle physics is incomplete. For example, if neutrinos have mass, what happened to all of the right-handed neutrinos? There are models that now include right-handed neutrinos, but since they have never been observed, these postulated right-handed neutrinos are referred to as 'sterile neutrinos'.

In the following sections, extensions to the Standard Model of particle physics are discussed, along with possible dark matter particle candidates. This chapter (and indeed the rest of this book) concentrates on four broad categories of dark matter particle type: WIMPs (weakly interacting massive particles), superWIMPs (superweakly interacting massive particles), sterile neutrinos and axions.

4.1 WIMPs

What solutions exist that could potentially solve the hierarchy problem? One solution is that of supersymmetry. Supersymmetry predicts a new class of particles called weakly interacting massive particles or WIMPs. They have predicted masses in the range 1.8×10^{-26}–1.8×10^{-24} kg (10 GeV–1 TeV). They can interact with W and Z gauge bosons, but not with gluons or photons. WIMPs are the most studied dark matter candidates and are found in many particle physics theories (not just supersymmetry).

If a WIMP exists and it is stable, it is naturally produced with a density in the Universe that is consistent with that expected for dark matter. This is referred to as the WIMP miracle. It is the main reason why WIMPs are so heavily studied.

Initially, the Universe was dense and hot, which meant that all particles were in thermal equilibrium (i.e., they all had the same temperature). It is important to realize at this point that energy is related to mass and temperature (see equations (2.2) and (3.1)). From a given amount of energy, mass can be created (equation (2.2)). Also, the energy (which is predominantly kinetic in the early Universe) can be equated to a temperature. Quite often in particle physics, energy, temperature and mass are all quoted in units of energy. The unit of choice is the electron-volt (eV) and 1 eV $= 1.6 \times 10^{-19}$ J. In the rest of the chapter, we talk about energy, mass, and temperature as if they are interchangeable (a common practice in particle physics) since they are all closely related to each other.

When the Universe cools to temperatures below the dark matter particle's mass, the number of dark matter particles drops exponentially as $e^{m_{DM}/T}$, where m_{DM} is the mass of the dark matter particle and T is the temperature of the Universe. Because WIMP candidates are self-annihilating (i.e., they are their own anti-particles), the number of particles would drop to zero. However, this does not happen because the Universe is expanding (as well as cooling), and eventually the Universe becomes so large and the number density of dark matter particles becomes so dilute that they cannot find each other. We say that the dark matter particles 'freeze out' and their number density asymptotically approaches a constant, which we refer to as their 'thermal relic density'.

This process is described quantitatively by the Boltzmann equation,

$$\frac{dn}{dt} = -3Hn - \langle \sigma_A v \rangle \left(n^2 - n_{eq}^2 \right) \tag{4.1}$$

where n is the number density of the dark matter particle, H is the Hubble parameter (the expansion rate of the Universe), $\langle \sigma_A v \rangle$ is the thermally averaged annihilation cross section, and n_{eq} is the dark matter number density in thermal equilibrium. The first term on the right hand side of equation (4.1), $-3Hn$, accounts for dilution from the expansion of the Universe. The n^2 term arises due to dark matter self-annihilation that results in a pair of Standard Model particles, and the n_{eq}^2 arises from the reverse process, i.e., a Standard Model pair annihilating to produce dark matter particles. The Boltzmann equation describes the statistical behavior of a thermodynamic system (in this case, the Universe), which is not in thermodynamic equilibrium. In the case of equation (4.1) it relates the rate of change of the dark matter particle density (the left hand side) to both the dilution due to the expansion of the Universe and the production and self-annihilation of dark matter particle pairs.

The thermal relic density can be determined by solving equation (4.1) numerically. If we define freeze out to be at the time when $n \langle \sigma_A v \rangle = H$, then the number density of dark matter particles at the time of freeze out is given by

$$n_f \approx \left(m_{DM} T_f \right)^{3/2} e^{-m_{DM}/T_f} \approx \frac{T_f^2}{M_{Pl} \langle \sigma_A v \rangle} \qquad (4.2)$$

where T_f is the background temperature at the time of freeze out and n_f is the number density of dark matter particles at the time of freeze out. We also define the ratio, $x_f \equiv m_{DM}/T_f$, which is in the exponential and is highly insensitive to the properties of the dark matter particle and may be considered a constant with a typical value of $x_f \approx 20$. The thermal relic density is then given by,

$$\Omega_{DM} = \frac{m_{DM} n_0}{\rho_c} = \frac{m_{DM} T_0^3}{\rho_c} \frac{n_0}{T_0^3} \approx \frac{m_{DM} T_0^3}{\rho_c} \frac{n_f}{T_f^3} \approx \frac{x_f T_0^3}{\rho_c M_{Pl}} \langle \sigma_A v \rangle^{-1} \qquad (4.3)$$

where the subscript 0 denotes present-day quantities and ρ_c is the critical density of the Universe. In equation (4.3), the parameter Ω_{DM} (called the dark matter 'density parameter') is the ratio of the density of dark matter to the critical density of the Universe. The thermal relic density does not depend on the mass of the dark matter particle, and it is inversely proportional to the annihilation cross section of the dark matter particle. On dimensional grounds, the cross section can be rewritten as

$$\sigma_A v = \frac{k g_{weak}^4}{16 \pi^2 m_{DM}^2} \qquad (4.4)$$

where the constant $g_{weak} \approx 0.65$ is called the weak interaction gauge coupling constant (this is essentially a constant that describes the strength of the weak nuclear force in this particular interaction), and k accounts for deviations from this estimate. The dark matter particle's mass is then given by

$$m_{DM} = \frac{g_{weak}^2}{4\pi} \sqrt{\frac{k \Omega_{DM} \rho_c M_{Pl}}{x_f T_0^3}} \qquad (4.5)$$

Given that we have good estimates for the quantities of the right hand side of equation (4.5), i.e., $\Omega_{DM} = 0.22$ (from WMAP), $T_0 = 2.73$ K (from COBE), the critical density $\rho_c = 9.47 \times 10^{-27}$ kg m^{-3}, and the Planck mass $M_{Pl} = 1.22 \times 10^{19}$ GeV (or 2.17×10^{-8} kg), we can determine that the typical mass for WIMPs is in the range $m_{DM} = 100$ GeV–1 TeV (with k in the range 0.5–2.0), if WIMPs are all of dark matter. The mass range we have derived here for WIMPs is in the weak scale (i.e., energy ranges which probe the weak nuclear force), and this coincidence is referred to as the WIMP miracle. This implies that many models of particle physics easily provide viable dark matter candidates, and it is the central reason why these problems in particle physics and astrophysics are related. As an aside, if WIMPs do not account for all of the dark matter, the mass range above would be somewhat lower.

For example, if it is shown that WIMPs only account for 10% of dark matter, then the mass range would be reduced accordingly, i.e., $m_{DM} = 10$–100 GeV.

There are indications that if new particles exist at the weak scale, at least one of them should be stable. The Large Electron Positron (LEP) Collider, which ran at CERN in the 1990s, was able to put limits on the dark matter particle in this manner (even though it was not designed to detect dark matter, the fact that it did not detect dark matter placed a lower limit on its mass range). This stable particle would not be able to decay. This gives us a candidate particle for dark matter in just the expected mass range, a mass range that reproduced the correct relic density for dark matter in the Universe.

There are several possible candidates for WIMPs. One of the most popular is the neutralino, a new particle, which comes out of supersymmetry. Supersymmetry appears to solve the hierarchy problem most elegantly, which is why it is highly favored by particle physicists. In supersymmetry, every Standard Model particle has a new, as-yet-undiscovered partner particle, which has the same quantum numbers and gauge interactions, but differs by spin 1/2. In other words, every Standard Model fermion, has a supersymmetric partner particle that is a boson. Also, every Standard Model boson has a supersymmetric partner that is a fermion. This doubling of the Standard Model particle spectrum has many implications for cosmology. For dark matter, we can start by listing all of the new particles that are electrically neutral. There is one spin-3/2 Fermion, the Gravitino, $\tilde{G}$. There are four spin-1/2 Fermions, which are denoted $\tilde{B}$, $\tilde{W}$, $\tilde{H}_u$, and $\tilde{H}_d$. These spin 1/2 Fermions mix to form four mass eigenstates, the neutralinos, χ_1, χ_2, χ_3, and χ_4. The lightest of these, χ_1, is a WIMP dark matter candidate. The gravitino is not a WIMP (it is actually a superWIMP which is a form of dark matter discussed in section 4.2), but it is a viable and fascinating dark matter candidate, as discussed later.

An alternative for new weak-scale physics is extra dimensions, which results in a type of dark matter candidate referred to as Kaluza–Klein dark matter. In these universal extra dimensions, all particles move in compact extra dimensions of size 10^{-18} m or smaller. In the simplest of these models, there is one extra dimension of size R and every Standard Model particle has an infinite number of partners, all of which have the same spin. These models do not solve the hierarchy problem, so the motivation for these models is not on the same strong ground as the supersymmetric models. Nevertheless, the lightest stable particles produced by Kaluza–Klein models are in the weak-scale range at around 1.0 TeV. Kaluza–Klein dark matter provides an instructive example of WIMPs that differ in important aspects from neutralinos.

4.2 SuperWIMPs

SuperWIMPs (or superweakly interacting massive particles) also have the required relic density for dark matter, but they have interactions that are much weaker than weak. SuperWIMP scenarios predict signals from cosmic rays and in astrophysics that are much more amenable to detection with experiments than WIMPs.

In this framework, dark matter is produced in late decays. WIMPs still freeze out as usual in the early Universe, but then decay later on to produce superWIMPs, which form the dark matter that exists today. SuperWIMPs are very weakly interacting, so they have no impact on WIMP freeze out, and WIMPs still decouple with the correct relic density for dark matter. Assuming that each WIMP decays to a superWIMP, the superWIMPs inherit the relic density of WIMPs.

The favored superWIMP candidate is the gravitino, $\tilde{G}$. Gravitinos are the 3/2-spin superpartners of gravitons (which is the gauge boson for the gravitational force in quantum gravity), and they exist in all supersymmetric models. The gravitino's mass is given by

$$m_{\tilde{G}} = \frac{F}{\sqrt{3}\, M_*} \tag{4.6}$$

where F is the square of the supersymmetric-breaking scale and $M_* = (8\pi G)^{-1/2} \approx 2.4 \times 10^{18}$ GeV is the reduced Planck mass. Supersymmetric breaking is the process to obtain a seemingly non-supersymmetric physics from a supersymmetric theory. This is a necessary step to reconcile supersymmetry with actual experiments. The supersymmetric breaking scale is the energy scale where supersymmetric breaking takes place. If supersymmetry fully resolves the hierarchy problem, this energy scale should not be too far from 1.0 TeV, and therefore should be accessible at the Large Hadron Collider at CERN.

In the simplest supersymmetric models the masses of supersymmetric partners are given by:

$$\tilde{m} \sim \frac{F}{M_*} \tag{4.7}$$

One particular solution to the hierarchy problem requires that $F \sim (10^{11}\ \text{GeV})^2$, and so all superpartners and the gravitino have weak-scale mass.

4.3 Sterile neutrinos

Neutrino oscillations indicate that neutrinos must have a mass. This requires new physics beyond the Standard Model of particle physics. In the Standard Model, all neutrinos are 'left-handed' and massless. The idea of particles being 'right-handed' or 'left-handed' is somewhat abstract, and the details will not be explained here. In a nutshell, and somewhat over-simplistically, whether a particle is 'right-handed' or 'left-handed' depends on whether its spin and direction of motion point in the same direction. Because neutrinos are all 'left-handed' the mechanism by which they obtain their mass must be complex, and it is definitely not the same mechanism that generates the masses for leptons and quarks. If we were to add 'right-handed' neutrinos, this problem would be solved. However, under the symmetries of the Standard Model, these right-handed neutrinos must have no Standard Model gauge interactions. These right-handed neutrinos are therefore referred to as 'sterile neutrinos'.

Sterile neutrinos may be produced in a number of ways, but unfortunately they do not naturally have the correct relic density. However, they may explain some experimental observations, which will be explained in chapter 6.

4.4 Axions

Axions are motivated by a problem in the Standard Model of particle physics that is referred to as the strong charge parity (CP) problem. This contributes to CP-violation and observables that conserve flavor, such as the electric dipole moment of the neutron, d_e. Quantum chromodynamics (QCD, the theory of strong nuclear interactions between quarks and gluons) does not break the CP-symmetry, although the theory indicates that there could be a violation of CP-symmetry in the strong interactions. CP-symmetry states that the laws of physics should be the same if a particle is interchanged with its antiparticle (C-symmetry), and then its spatial coordinates are inverted ('mirror' of P-symmetry). Violation of CP-symmetry was first discovered in 1964 via the decay of neutral kaons, but this decay is a result of the weak nuclear interaction. CP-violation is critical in cosmology, as it is the phenomenon that generates the observed matter–antimatter asymmetry of the Universe. Incidentally, kaons, which are sometimes called K mesons and are denoted with the symbol K, are any of a group of four mesons distinguished by a quantum number called strangeness (which is related to the strange quark). In the quark model they are understood to be bound states of a strange quark (or antiquark) and an up or down antiquark (or quark). Mesons are subatomic particles that always consist of a quark and an antiquark. They are one of two mechanisms in particle physics for binding quarks together. The other type is called a baryon, in which three quarks are bound together. For instance, the proton is a baryon that consists of two up quarks and one down quark. Collectively, baryons and mesons are referred to as hadrons, i.e., hadrons are particles that consist of quarks.

In QCD and strong interactions, there is no experimentally known violation of CP-symmetry. For example, the neutron electric dipole moment (d_e) has never been observed experimentally, but current constraints imply that $d_e < 2.9 \times 10^{-26}$ e cm. This is therefore a fine-tuning problem of 1 part in 10^{10}, and this is the motivation for axions as dark matter candidates.

The allowed parameters for axions imply that they are very light and weakly interacting. This provides another qualitatively different dark matter candidate that is well motivated by particle physics. The mass and interactions of the axion are defined by the axion decay constant, f_a, which is a new mass scale. The mass of the axion is then given by

$$m_a = \frac{\sqrt{m_u m_d}}{m_u + m_d} m_\pi f_\pi \frac{a}{f_a} \approx 6\,\mu\text{eV}\left(\frac{10^{12}\,\text{GeV}}{f_a}\right) \tag{4.8}$$

where $m_u \approx 4$ MeV is the mass of the up quark, $m_d \approx 8$ MeV is the mass of the down quark, and $m_\pi \approx 135$ MeV is the pion mass, and $f_\pi \approx 93$ MeV is the pion decay constant.

The mass of the axion is dependent upon several constraints. In magnetic fields, axions are predicted to change to and from photons. This coupling of the axion to the photon results in a decay time for the axion that is inversely proportional to its mass to the fifth power, m_a^5. This means that for axions to live longer than the age of the Universe, the axion mass must be tiny, $m_a \leqslant 20$ eV. Other astrophysical constraints result in even smaller masses for the axion. For instance, axions should be created in evolved stars (due to the interaction of photons with the stellar magnetic field), but the longevity of red giants and the observed length of the neutrino outburst from Supernova 1987A require the axion decay constant to be $f_a \geqslant 10^9$ GeV, which implies an axion mass of $m_a \leqslant 10$ eV.

There are several production mechanisms for axion dark matter. The most straightforward is thermal production, which leads to a relic density of

$$\Omega_a \sim 0.22\left(\frac{m_a}{80 \text{ eV}}\right) \tag{4.9}$$

but this would be hot dark matter. Also, equation (4.8) results in lifetimes shorter than the age of the Universe for 80 eV axions. Therefore, thermal production cannot produce axions that are the bulk of the dark matter. For the limit we determined from Supernova 1987A, where the axion mass limit was found to be $m_a \leqslant 10$ eV, axions would only account for 1/8 of the dark matter in the Universe. In other words, the relic density, as calculated from equation (4.9), would be only $\Omega_a \approx 0.028$.

There are various non-thermal production mechanisms for axions, which result in a relic density of

$$\Omega_a \approx 0.4\theta_i^2\left(\frac{f_a}{10^{12} \text{ GeV}}\right)^{1.18} \tag{4.10}$$

where θ_i is an arbitrary constant less than or equal to 1. Given that from **WMAP** results, we know that the density of axions, $\Omega_a \leqslant \Omega_{DM} = 0.22$, equation (4.10) implies that the axion decay constant, $f_a \leqslant 10^{12}$ GeVθ_i^{-2}. Depending on when inflation occurs, we can also put further bounds on the axion decay constant, f_a, and the axion mass, m_a, such that

$$10^{12} \text{ GeV}\theta_i^2 \geqslant f_a \geqslant 10^9 \text{ GeV}$$
$$6 \text{ }\mu\text{eV}\theta_i^2 \leqslant m_a \leqslant 6 \text{ meV} \tag{4.11}$$

and if the constant $\theta_i \sim 1$, then axion production implies a slightly stronger upper bound on the axion decay constant, f_a.

The lower bound on the axion mass in equation (4.11) arises by requiring that axions do not over-close the Universe (after all, our observations of the cosmic microwave background by **WMAP** and other experiments indicate that the Universe is very close to flat). If this lower bound is saturated, axions should be all of dark matter, and this is therefore the target for dark matter searches. It should be noted that axions do not naturally have the correct relic density because there is a range of allowed masses for the axion.

There are some other forms of proposed dark matter, but for the purposed of this book, we will stop here. In the next two chapters, we will discuss the possible experiments and methods that can be used (and in some cases are being used) to detect the various types of dark matter. This can include particle colliders, astrophysical detection, and direct detection.

Suggested further reading

Cottingham W N and Greenwood D A 2007 *An Introduction to Modern Particle Physics* (Cambridge University Press: Cambridge)

Feng J L 2010 *Annu. Rev. Astron. Astrophys.* **48** 495–545

Guralnik G S, Hagen C R and Kibble T W 1964 *Phys. Rev. Lett.* **13** 585–87

Higgs P W 1964 *Phys. Lett.* **12** 132–33

Higgs P W 1964 *Phys. Rev. Lett.* **13** 508–9

Kane G 2013 *Supersymmetry and Beyond: From the Higgs Boson to the New Physics* (Basic Books: New York)

Kuster M, Raffelt G and Beltran B 2007 *Axions: Theory, Cosmology, and Experimental Searchers* (Springer: Berlin)

Smirnov A Y and Zukanovich Funchal R 2006 *Phys. Rev.* D **74** 013001

Chapter 5

Indirect detection of dark matter

In this chapter, we will learn how dark matter can be detected indirectly (direct detection of dark matter will be discussed in chapter 6). The exact methods that can be used for indirect detection, once again, depend on the type of dark matter particle we are trying to detect. This chapter summarizes these ideas.

In most indirect detection scenarios, scientists make use of the prediction that most dark matter candidates are Majorana fermions. They are their own anti-particles. WIMP and superWIMP dark matter candidates are always Majorana fermions. As of yet, it is unclear if neutrinos (and therefore sterile neutrinos) are Majorana fermions or regular (Dirac) fermions. Some of the methods for indirect detection of dark matter, described in this chapter, rely on this property of dark matter, and they essentially look for the signals of the resulting decay products when the dark matter self annihilates.

5.1 WIMPs

After freeze out, dark matter pair annihilation becomes greatly suppressed. However, dark matter annihilation does still continue and should have continued to this day, so it should be observable. As a result, it may be possible to detect WIMP dark matter indirectly. If a dark matter pair annihilates, it could produce something that could somehow be detected (such as some combination of Standard Model particles). There are many indirect detection methods that are being considered for WIMPs.

One possible method of indirect detection comes from neutrino searches. When WIMPs pass through the Sun or the Earth, they may scatter and be slowed enough to be captured by the gravitational potential of the Sun or Earth. They can then settle to the center, where their particle densities and annihilation rates are significantly enhanced. Most of their annihilation products are immediately absorbed, except one: neutrinos. Some of these resulting neutrinos can travel to the surface of the Earth, where they can be converted to charged leptons, which in turn can be detected.

The neutrino flux depends upon the WIMP density, which is determined by the competing processes of gravitational capture and pair annihilation. It turns out that under fairly general conditions, neutrino searches are directly comparable to direct detection. Experiments such as Super-Kamiokande[1], the IceCube Neutrino Observatory[2], the Antarctic Muon and Neutrino Detector Array[3] (AMANDA), and the Astronomy with a Neutrino Telescope and Abyss environmental RESearch project[4] (ANTARES) have searched for excesses of neutrinos from the Sun with energies in the range $10\,\mathrm{GeV} \leqslant E_\nu \leqslant 1\,\mathrm{TeV}$. Null results from these experiments provide the leading bounds on spin-dependent cross sections for WIMPs. These detectors are starting to probe the relevant regions of supersymmetric parameter space. Neutrino searches can also be sensitive to spin-independent cross sections. Future neutrino searches at Super-Kamiokande and ANTARES may have lower thresholds and may thus provide bounds on low-mass WIMPs, which may therefore test the DAMA signal regions (see chapter 6) with high sensitivity and in the mass range $m_{\mathrm{DM}} \sim 1$–$10\,\mathrm{GeV}$.

In addition to neutrinos, there are many other particles that may be signals of dark matter annihilation. There have been many reported anomalies in indirect detection, a lot of which have been attributed to dark matter. A recent example is the detection of positrons and electrons with energies between 10 GeV and 1 TeV by the Payload for Antimatter/Matter Exploration and Light-nuclei Astrophysics[5] (PAMELA), Advanced Thin Ionization Calorimeter[6] (ATIC), and Fermi Large Area Telescope[7] (Fermi-LAT) collaborations (see figure 5.1 for the Fermi-LAT gamma-ray map of the sky). The data from these experiments reveal an excess signal above the modeled background. There are some plausible astrophysical explanations for this excess. The ATIC and Fermi experiments cannot distinguish electrons from positrons, and so they constrain the total electron + positron flux. Also, the excess seen in the ATIC data is not confirmed by Fermi-LAT, which has much higher statistics. Furthermore, the Fermi data can be explained by a modification of the spectral index of the cosmic ray background. Finally, the PAMELA data is actually consistent with the expected backgrounds from pulsars, which are rapidly rotating neutron stars, and end product of stellar evolution. Nevertheless, it is still interesting to explore the possibility that the positron excesses are a result of dark matter annihilation, and this is the argument we will now follow.

The energies of the excess, which are around m_{weak}, are just as expected for WIMPs. However, the observed fluxes are much larger than expected for WIMPs. In order for WIMP annihilation to have the correct thermal relic density, the WIMP annihilation cross-section must be $\sigma \equiv \langle \sigma_A v \rangle \approx 3 \times 10^{-26}\,\mathrm{cm}^3\,\mathrm{s}^{-1}$.

---

[1] http://www-sk.icrr.u-tokyo.ac.jp/sk/index-e.html

[2] https://icecube.wisc.edu/

[3] http://amanda.uci.edu/

[4] http://antares.in2p3.fr/

[5] http://pamela.roma2.infn.it/index.php

[6] http://atic.phys.lsu.edu/

[7] https://www-glast.stanford.edu/

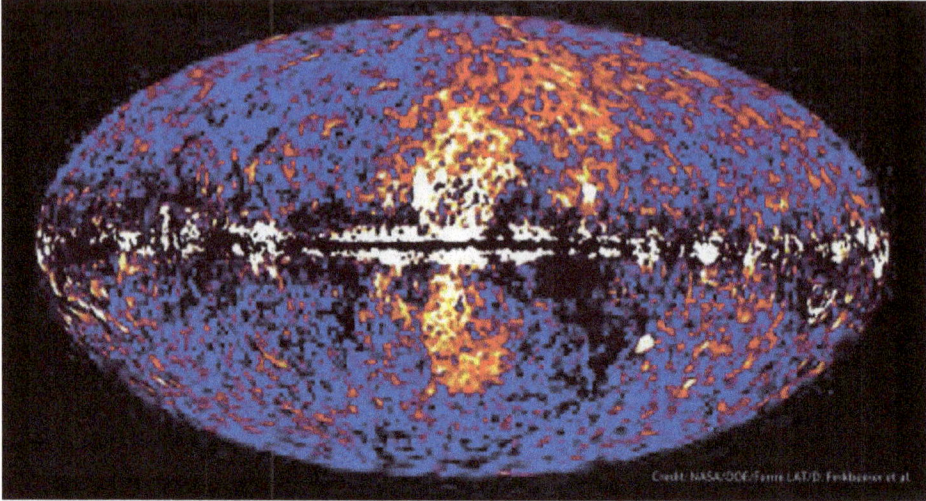

Figure 5.1. Residual map of the Fermi-LAT (Large Area Telescope) gamma-ray sky in galactic coordinates after subtraction of known point sources and diffuse emission. Two lobe-like structures aligned with the Galactic Center and extending to more than 50 degrees in galactic latitude are apparent in these residuals. The relationship of these structures as well as the gamma-ray emission from the galactic center to the expected signatures of dark matter annihilations is currently an active area of investigation. Image credits: NASA/ Fermi-LAT experiment.

This must be 100–1000 times larger to explain the positron data. Astrophysical arguments from substructure cannot explain these enhancements. There are, however, particle physics arguments that can explain the enhancement. For instance, let us assume that the dark matter interacts with a hypothetical light force carrier (ϕ) with a fine structure constant $\sigma_{DM} \equiv \lambda^2/(4\pi)$. If the force carrier, ϕ, is massless, its cross section can be enhanced by the Sommerfeld enhancement factor, i.e.,

$$S = \frac{\pi\alpha_{DM}/v_{rel}}{1 - e^{-\pi\alpha_{DM}/v_{rel}}} \tag{5.1}$$

which is an effect that was first derived for electron positron annihilation. If the force carrier, ϕ, has mass, S (the Sommerfeld enhancement factor) is cut off at a value, which is proportional to $\sigma_{DM} m_{DM}/m_\phi$. At freeze out, the relative velocity of dark matter particles is $v_{rel} \sim c/3$, and now, at the present time, it is about $v_{rel} \sim 10^{-3}c$. This Sommerfeld enhancement therefore provides a rather elegant boosting mechanism that could potentially explain the observed annihilations. Indeed, Sommerfeld enhancement alone can explain the excesses seen in the PAMELA and Fermi data.

In addition to neutrinos from the Sun and positrons from the galactic halo, there are some other promising indirect search strategies. Searches for antiprotons and antideuterons (which are a combination of an antiproton and antineutron) from WIMP annihilation in the galactic halo can provide complementary searches.

Searches for gamma rays by space-based experiments, such as Fermi and AMS, and by ground-based atmospheric Cherenkov telescopes are also promising. Gamma rays are produced when WIMPs annihilate to other particles, which then radiate photons, leading to a smooth distribution of gamma-ray energies. However, photons point back to their source, and so they can provide a very powerful diagnostic tool. Targets for gamma-ray sources are the Galactic center, where signal rates are high but background rates are also high and potentially hard to estimate, and dwarf galaxies, where signal rates are lower, but backgrounds are also expected to be low.

5.2 SuperWIMPs

SuperWIMPs cannot be detected directly, they are simply too weakly interacting and their annihilation signal rates are completely negligible. However, if the decaying WIMP is charged, the superWIMP scenario implies long-lived charged particles. One possibility is that long-lived charged particles can be produced by high-energy cosmic rays, which would result in exotic signals in cosmic ray and neutrino experiments. For example, in the gravitino superWIMP case with a stau (the supersymmetric partner for the tau lepton) as the next lightest supersymmetric particle (NLSP), ultrahigh-energy neutrinos could produce events with two long-lived staus. These metastable staus can propagate to neutrino detectors, and they could therefore be detected as events with two extremely high-energy charged tracks in experiments such as IceCube.

Because superWIMPs are very weakly interacting, the decays of WIMPs into superWIMPs may occur very late and therefore could have an observable effect on Big Bang nucleosynthesis and the cosmic microwave background. Additionally, superWIMP dark matter may behave as warm dark matter. When WIMPs decay into superWIMPs, the accompanying particles may distort the frequency dependence of the cosmic microwave background (CMB) away from its ideal black body spectrum.

Late time energy release after $t \sim 1$ s also destroys and creates light elements, and could thus potentially distort the predictions of standard Big Bang nucleosynthesis (BBN). This depends on the exact nature of the NLSP. In the neutralino NLSP case, there are possible decay modes that could lead to hadrons. However, constraints from BBN on hadronic energy release exclude the neutralino as an NLSP WIMP. In the charged slepton NLSP scenario (e.g., the stau), the decaying WIMP could bind with atomic nuclei, which would enhance the effect of its decays on BBN. Late decays to superWIMPs could improve the current disagreement of BBN predictions with observed lithium-6 and lithium-7 abundances, although this typically requires that the decaying slepton be heavy, with a predicted mass above 1 TeV.

SuperWIMPs are produced with large velocities at late times. This means that the resulting velocity dispersion would reduce the phase space density, smoothing out cusps in dark matter halos and potentially creating cores. Also, such particles damp the linear power spectrum, reducing power on small scales. SuperWIMPs can suppress small-scale structure and behave just like warm dark matter. Some superWIMP scenarios may therefore be differentiated from standard cold dark matter scenarios by their impact on small-scale structure. It is important to mention that, although superWIMP dark matter can create cores, it is still a cold dark matter candidate (unlike the sterile neutrino, which is, in fact, a warm dark matter candidate).

5.3 Sterile neutrinos

The dominant decay of sterile neutrinos is given by

$$\nu_S \rightarrow \nu_L \bar{\nu}_L \nu_L$$

where ν_S is a sterile neutrino, ν_L are both left-handed neutrinos, and $\bar{\nu}_L$ is a left-handed antineutrino. However, sterile neutrinos can also decay through a loop-level process to a photon and an active neutrino. The branching ratio (i.e., the fraction of particles which decay in this latter manner with respect to the total number of sterile neutrinos that decay) is about 1/128. For the allowed sterile neutrino parameters, the lifetime of a sterile neutrino is much longer than the age of the Universe, which is required for it to be dark matter.

Even if a small number of sterile neutrinos decay, this could be observed. The radiative decay is two-body so the signal would be a mono-energetic flux of x-rays with energy $E_\gamma = m_S/2$ where m_S is the mass of a sterile neutrino in eV. Such signals could potentially be seen by space-based x-ray observatories, such as Chandra or XMM-Newton. There is reported evidence of a signal in Chandra data, consistent with $(m_S, \sin^2 2\theta) \sim (5 \text{ keV}, 3 \times 10^{-9})$, in the heart of the allowed parameter space. Future observations from the European Space Agency's proposed ATHENA X-ray observatory[8] may help extend the sampled parameter space.

Sterile neutrinos are a classic warm dark matter candidate. However, their temperature depends on how they are produced. The sterile neutrino free-streaming length is given by

$$\lambda_{FS} \approx R \frac{\text{keV}}{m_S} \tag{5.2}$$

where $R = 0.2\text{--}0.9$ Mpc depending on the production mechanism. This implies that bounds on small-scale structure (e.g., through constraints from Lyman-α observations) depend on the production mechanism. For production by neutrino oscillations, current bounds require a sterile neutrino mass of $m_S \geqslant 10 \text{ keV}$, thus excluding this mechanism as a source for all of dark matter. Lyman-α bounds are weaker, which suggests that colder production mechanisms could viably produce all of the dark matter. In addition to their impact on small-scale structure and the x-ray spectrum, sterile neutrinos may have other astrophysical effects on the velocity distribution of pulsars and on the formation of the first stars.

Suggested further reading

Abassi R *et al* 2012 *Phys. Rev.* D **85** 042002

Albert A 2015 *Searching for Dark Matter with Cosmic Gamma Rays* (Bristol: IOP Publishing)

---

[8] http://www.the-athena-x-ray-observatory.eu

ANTARES Collaboration 2013 *J. Cosmol. Astropart. Phys.* **11** 032

Cho K *et al* 2015 *Phys. Rev. Lett.* **114** 141301

El Zant A A, Khalil S and Okada H 2010 *Phys. Rev.* D **81** 123507

Esmali A, Kang S K and Dario Serpico P 2014 *J. Cosmol. Astropart. Phys.* **12** 054

Kong K and Park J-C 2014 *Nucl. Phys.* B **888** 154–68

Chapter 6

Direct detection of dark matter

This chapter summarizes the techniques that are used to directly detect dark matter. Not all dark matter candidates can be detected directly (as described in chapter 5, specifically for superWIMPs). Here, we will learn about the techniques that are being used to detect certain types of dark matter.

6.1 WIMPs

One of the easiest forms of proposed dark matter candidates that could be directly detected are WIMPs. WIMPs may be detected by their scattering off normal matter. Given a typical WIMP mass of $m_{DM} \sim 100$ GeV and WIMP velocity of $v \sim 300$ km s^{-1}, the deposited recoil energy is at most ~ 100 keV, requiring highly sensitive, low-background detectors placed deep underground. Deep underground detectors probe the weak cross section frontier, which includes WIMPs (as well as other rare processes in physics).

The field of direct WIMP detection is extremely active. Sensitivities for these detectors have increased by a factor of about 100 in the last decade and they are continually improving. One of the newest dark matter detectors is DarkSide[1], a liquid argon detector at the Gran Sasso National Laboratory (LNGS) in Italy (see figure 6.1). Another detector, also in Gran Sasso is the DAMA[2] (Dark Matter Experiment) liquid xenon (LXe) detector (DAMA actually consists of several experiments, but we will focus on LXe for now), and also the XENON[3] detector and Large Underground Xenon (LUX)[4] dark matter detector (which are both also liquid xenon experiments). Argon and xenon are both noble gases. In both of these experiments, the noble gas, whether argon or xenon, is in liquid form. Both types of

---

[1] http://darkside.lngs.infn.it/ds-50/
[2] http://people.roma2.infn.it/~dama/web/home.html
[3] http://xenon.astro.columbia.edu/
[4] http://lux.brown.edu/LUX_dark_matter/Home.html

doi:10.1088/978-1-6817-4118-5ch6

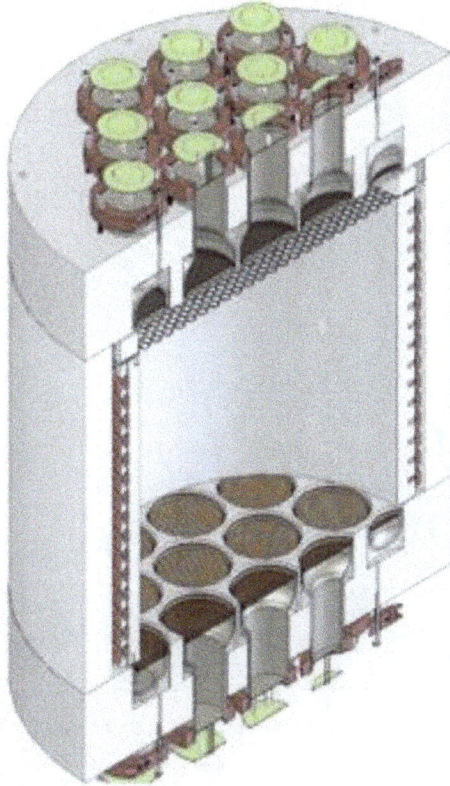

Figure 6.1. Schematic of the DarkSide-50 liquid argon dark matter detector, which is stationed deep underground in Hall C at the Gran Sasso National Laboratory in Italy.

experiments make use of spin-independent scattering. In this case, a WIMP enters the active part of the detector (which contains argon or xenon in liquid form) and strikes an atomic nucleus. The energy of the WIMP (or at least some of its kinetic energy) makes the atomic nucleus recoil, and as a result a photon (with a frequency that is equivalent to the recoil energy, $f = E/h$, where E is the recoil energy, and $h = 6.626 \times 10^{-34}$ m^2 kg s^{-1} is Planck's constant) is emitted and can be detected by photomultiplier tubes (PMTs), as shown in figure 6.2.

One of the biggest problems for this type of detector is that of background signals. Most of the background signals can be eliminated by putting the detectors deep underground. However, some cosmic rays, especially relativistic muons can still penetrate the rock and potentially make it to the detector. This is not such a big deal, since muons carry an electric charge and can easily be distinguished from neutral massive WIMPs. However, muons can interact with the rock, and this may lead to neutrons entering the detector. Since neutrons are neutral, if they make it to the active part of the detector, the signal they could potentially create would be difficult to distinguish from the signal we would expect for a WIMP. These neutrons are called cosmogenic neutrons. The materials making up the detector will almost certainly be radioactive, albeit at very low levels. However, when the goal is to detect

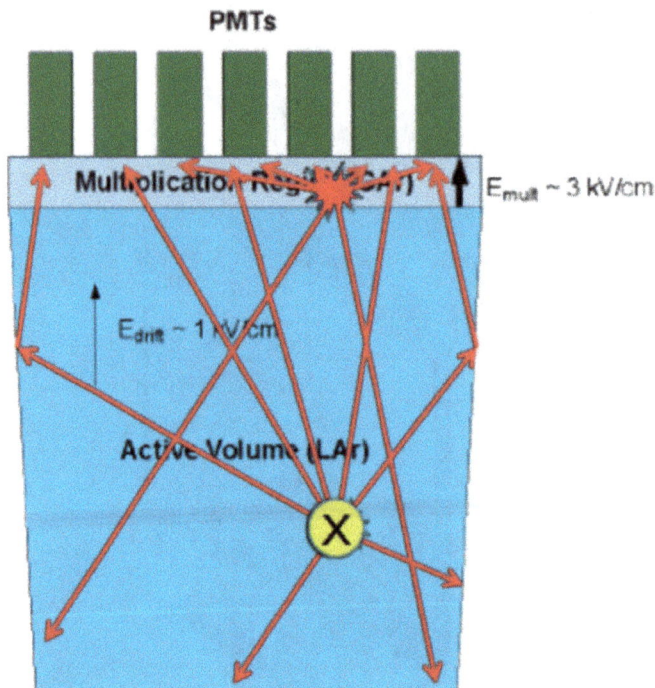

PMTs

Multiplication Region (GAr) $E_{mult} \sim 3\,kV/cm$

$E_{drift} \sim 1\,kV/cm$

Active Volume (LAr)

X

Figure 6.2. Spin-independent scattering of dark matter particles, in this case for a liquid argon detector, such as DarkSide.

rare events, such as dark matter interactions, even low-level radioactivity is undesirable. One of the byproducts of this low-level radioactivity will be radiogenic neutrons, which could also potentially be difficult to distinguish from WIMPs if they were detected by the active detector.

So what is the solution to this problem? This is where active suppression systems come into play. Radiogenic neutrons have relatively low energies and hence do not penetrate far through most materials. On the other hand, they are produced inside the materials that make up the detector by the decays of long-lived radio-isotopes in those materials. All direct detection dark matter experiments minimize radiogenic neutron production by taking pains to select detector construction materials with very low levels of intrinsic radioactivity. Some radiogenic neutron production is unavoidable, however, so to further suppress this background the dark matter detectors are deployed within a liquid scintillator veto (LSV), which is usually a tank filled with some kind of liquid scintillator. Neutrons that are produced in the dark matter detector are captured with high efficiency by the liquid scintillator (either before or after interacting in the active part of the detector), which produces a signal in the neutron veto. Thus, dark matter detectors can suppress the rate of background events from neutrons by using the coincidences between the neutron veto and the inner detector to reject radiogenic neutron recoil background events in the active detector.

Cosmogenic neutrons have higher energies than radiogenic neutrons, and can therefore penetrate much further through matter. Locating the dark matter

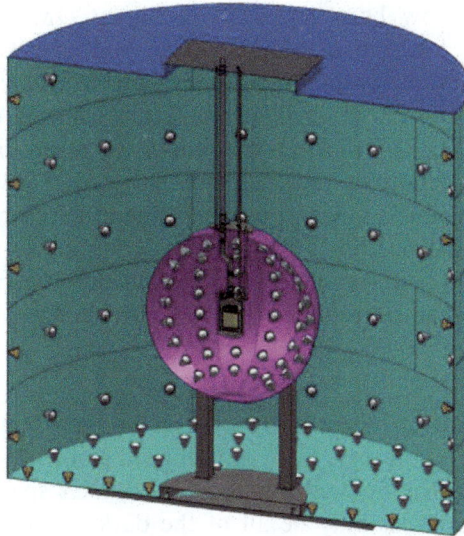

Figure 6.3. Sketch of an active detector immersed in a spherical liquid scintillator veto, which is in turn surrounded by a large water tank. This example is based upon the DarkSide-50 set up, which uses the Borexino[5] CTF (Counting Test Facility) water tank (the Borexino experiment was actually a neutrino detector which ran from 2007 until 2014).

experiments underground reduces the number of cosmogenic neutrons by reducing the rate of cosmic ray muons. The neutron veto will also be effective against cosmogenic neutrons, although their higher energy makes them more likely to penetrate the veto without leaving a signal. In order to more fully suppress cosmogenic neutron backgrounds, therefore, the liquid scintillator veto, containing the active detector, is usually deployed in a large water tank as shown in figure 6.3. Photomultiplier tubes in the water tank can then be used to detect the Cerenkov light produced as the muons (and other particles, including neutrons, which result from the muon shower) traverse the water. Large water tanks are chosen so that particles (particularly cosmogenic neutrons) from cosmogenic showers can be vetoed with a high efficiency.

For these spin-independent scattering detectors, there are both observed signals from the DAMA collaboration and null results from many other experiments. DAMA continues to find a signal in annual modulation with a period and maximum at the expected values for WIMP dark matter. From the theory side, the DAMA results are quite puzzling, because the signal seemingly implies dark matter masses and scattering cross sections that have been excluded by other experiments.

Spin-dependent scattering provides an independent method to search for dark matter. Probably the leading direct detection experiment is the Cryogenic Dark Matter Search[6] (CDMS) at the Soudan Underground Laboratory[7] in Northern

---

[5] http://borex.lngs.infn.it/

[6] http://cdms.berkeley.edu/

[7] http://www.sudan.umn.edu/

Minnesota. This has now been replaced by SuperCDMS[15], and will soon be relocated to a deeper location at SNOLAB[8] in Sudbury, Canada (the name of this location comes from the original Sudbury Neutrino Observatory, or SNO, that was located there). SuperCDMS will increase the chances of having a positive identification of a WIMP or will allow much more stringent limits to be placed on the interaction cross-section of these elusive particles. However, these types of experiments are less promising than spin-independent detectors in terms of probing the heart of supersymmetric and WIMP parameter space. In addition, given some reasonable assumptions, indirect detection experiments that are looking for dark matter annihilation to neutrinos in the Sun can provide more stringent constraints.

6.2 Axions

Axions may also be detected directly by looking for their scattering with Standard Model particles in the laboratory. For cosmological axions, given all of the caveats discussed in chapter 4, the favored region of parameter space, in particular their mass range (in which axions may be all of the dark matter), can be taken to be $1 \ \mu eV \leqslant m_a \leqslant 100 \ \mu eV$. In this mass range, the leading experiment is the Axion Dark Matter experiment [9] (ADMX). ADMX searches for cosmological axions by looking for the resonantly enhanced conversion of dark matter axions to photons through scattering off of a background magnetic field. The experiment detects the very weak conversion of dark matter axions into microwave photons. This axion-to-photon conversion process is stimulated by an apparatus consisting of an 8 tesla magnet and a cryogenically cooled tunable microwave cavity. The idea here is to tune the resonant frequency of the cavity to the axion mass. In this case, we have to understand the nature of wave–particle duality. At the quantum level, particles (such as the axion) have properties that can be more easily described if we treat them as waves (for instance, they appear to have a frequency and wavelength). The opposite can also be true. On a macroscopic-level, for instance, it is easiest to describe light as a wave, but at the quantum level we talk about the photon, which is essentially a particle description of light. So, at the quantum level, the axion can be described as having a frequency, and this is related to the energy (or mass if we remember that $E = mc^2$) via the following relationship

$$E = hf \tag{6.1}$$

where E is the energy of the particle (its mass-energy), f is its frequency, and $h = 6.626 \times 10^{-34} \ m^2 \ kg \ s^{-1}$ is Planck's constant. This allows experiments like ADMX to tune the resonant frequency of the cavity to the mass of the axion (or its equivalent energy, which is why we always measure masses of particles in eV, a unit of energy). When the resonant frequency of the cavity is tuned to the axion mass, the interaction between nearby axions and ADMX's magnetic field is enhanced. This results in the deposit of a very tiny amount of power (less than

---

[8] https://www.snolab.ca/
[9] http://depts.washington.edu/admx/home.html

10^{-24} watts) into the cavity. An extraordinarily sensitive microwave receiver allows the very weak axion signal to be extracted from the noise. The experiment's receiver features quantum-limited noise performance delivered by an exotic superconducting quantum interference device (SQUID) amplifier and lower temperatures from a liquid helium-3 refrigerator. ADMX is the first experiment sensitive to realistic dark-matter axion masses and couplings and the improved detector allows an even more sensitive search. ADMX has already eliminated one of the two axion benchmark models from 1.9 μeV to 3.53 μeV, assuming axions saturate the Milky Way's halo. ADMX hopes to exclude or discover 2–20 μeV dark matter axions within the next 10 years. ADMX is also currently undergoing an upgrade to what is being referred to as the 'definitive experiment'; this is sensitive to a very broad range of plausible dark-matter axion masses and couplings. Greater sensitivity will be possible with the upgrade to SQUID amplifiers.

6.3 Detection of WIMPs and superWIMPs in particle colliders

WIMPs and SuperWIMPs can potentially be detected in particle colliders, such as the Large Hadron Collider[10] (LHC; see figures 6.4 and 6.5) at CERN[11]. Given the energy of the LHC and the requirement that WIMPs have masses in the weak range and interact through the weak nuclear force, WIMPs will almost certainly be created at the LHC. Unfortunately, direct WIMP production of $\chi\chi$ pairs (e.g., neutralino pairs) is invisible. Searches for dark matter at the LHC, therefore, rely on indirect

Figure 6.4. A section of the Large Hadron Collider (LHC) at CERN in Geneva, Switzerland. Image credits: CERN.

---

[10] http://home.web.cern.ch/topics/large-hadron-collider/
[11] http://home.web.cern.ch/

production, although we will discuss it here so as not confuse this detection with the more astrophysical types of indirect experimentation. For example, in super-symmetry, the LHC should produce pairs of squarks (the supersymmetric partners to quarks) and gluinos (the supersymmetric partners to gluons). These particles could then decay through some cascade chain, and they would eventually end up as neutralino WIMPs, which escape the detector. Their existence is registered through the signature of missing energy and momentum, a signal that is a standard tool in searches for physics beyond the Standard Model of particle physics. Observations of missing particles may appear consistent with the production of dark matter, but it is far from compelling evidence. Indeed, such an observation implies that a particle was produced that was stable enough to exit the detector, typically implying a lifetime of $\tau \geqslant 10^{-7}$ s. However, dark matter lifetimes have to be at least as long as the age of the Universe or much older. Typically, dark matter models require lifetimes of $\tau \geqslant 10^{17}$ s and this is based upon cosmological arguments. The limits of lifetimes that can be placed on undetected particles in collider experiments is a far cry from the limits imposed on dark matter particle candidates by cosmological arguments.

Luckily, in the past few years, there has been a great deal of progress in this direction. Colliders, such as LHC, can perform detailed studies of new physics, and this can constrain the dark matter candidate's properties so strongly that its thermal relic density can be precisely determined. The consistency of this density with the cosmologically observed density would then provide more compelling evidence that the particle produced at the collider is the cosmological dark matter. Physics experiments at the LHC may stringently constrain cross sections involving dark matter and other particles. Along with knowledge of a cooling and expanding Universe, these microscopic data allow us to determine the dark matter relic density. This can then be compared with the observed density of dark matter, which would then give us confidence that dark matter was produced in the collider.

It should be noted that in its initial run, reaching energies of up to 7 TeV, the LHC found no evidence of supersymmetric particles. This is potentially a problem

Figure 6.5. An aerial photograph of CERN. The large circle indicates the Large Hadron Collider (LHC), with a circumference of 27 km (17 miles). The photograph depicts the position of four particle-detector experiments (ALICE, ATLAS, CMS, and LHCb) constructed at the LHC. Image credits courtesy of CERN/LHC.

for supersymmetry as a model for particle physics beyond the Standard Model, and it is therefore a potential problem for the neutralino as a candidate for dark matter (although it does not rule out other types of WIMPs, such as Kaluza–Klein dark matter). Nevertheless, with the LHC due to start running again, this time reaching higher energies of 14 TeV, it is still possible that the LHC finds evidence for supersymmetry. Indeed, some theorists believe that these higher energies need to be probed to really see the effects predicted by supersymmetric models of particle physics.

Particle colliders may also find evidence of superWIMPs. This evidence may come in one of two forms. Collider experiments may see long-lived, charged particles. Given the stringent bounds on charged dark matter, such particles should decay quickly, and their decay products should be superWIMP dark matter. Alternatively, colliders may find seemingly stable particles, but precision studies may find that these particles have a relic density that is too large. These two possibilities are not mutually exclusive, and the discovery of long-lived, charged particles with predicted relic densities that are too large is a distinct possibility, which would strongly motivate superWIMP dark matter.

Consistency between the particle physics predictions and the cosmological observations would provide compelling evidence that the particle produced at the LHC is, in fact, dark matter. Along the way, colliders can also determine the dark matter particle's mass, spin, and many other properties. In this way, colliders may finally help solve the question of the microscopic identity of dark matter. Such dark matter studies should provide a window on the era of dark matter freeze out at 1 ns (10^{-9} s) after the Big Bang and temperatures of ~10 GeV. Of course, the thermal relic density prediction from colliders and the cosmological observations need not be consistent. In this case, there are many possible lines of inquiry, depending on which is larger.

Now that the LHC has discovered the existence of the Higgs Boson, and with the LHC coming online again in 2015, experiments at CERN are likely to start looking for dark matter candidates with more vigor and interest. So watch this space.

Suggested further reading

Agnes P et al 2015 Phys. Lett. B **743** 456–66

Aprile E et al 2011 Phys. Rev. Lett. **107** 131302

Alexander T et al 2013 J. Instrum. **8** C11021

Alexander T et al 2013 Astropart. Phys. **49** 44–51

Busoni G, De Simone A, Jacques T, Morgante E and Riotti A 2015 J. Cosmol. Astropart. Phys. **03** 022

Cota R C, Rajaraman A, Tait T M P and Wijangco A M 2014 Phys. Rev. D **90** 013020

Wagner A et al 2010 Phys. Rev. Lett. **105** 171801

Chapter 7

What the future holds

The Planck satellite[1] and the Wilkinson Microwave Anisotropy Probe[2] (WMAP) have both revealed the so-called 'energy budget' of the Universe, and this is shown in figure 7.1 (for the Planck data). In this energy budget, matter (regular matter, i.e., atoms, plus dark matter) accounts for 31.7% of this budget, with the remaining 68.3% coming in the form of dark energy. The first thing we can conclude from this is that, even if we learn the nature of dark matter in the next decade, we still have no idea what most of the Universe is made of. Dark energy (which is in no way related to dark matter) is even more mysterious than dark matter. Dark energy is the hypothesized source of energy that is accelerating the expansion rate of the Universe. Many things about the nature of dark energy remain matters of speculation. The evidence for dark energy is indirect but comes from three independent sources:

- Distance measurements using Supernovae Type 1a[3] and their relation to redshift, which suggest the Universe has expanded more in the last half of its life (see figure 7.2).
- The theoretical need for a type of additional energy that is not matter or dark matter to form the observationally flat Universe (absence of any detectable global curvature).
- It can be inferred from measures of large-scale wave-patterns of mass density in the Universe.

Dark energy is thought to be very homogeneous, not very dense and is not known to interact through any of the fundamental forces other than gravity. Since it is quite rarefied—roughly 10^{-30} g cm^{-3}—it is unlikely to be detectable

---

[1] http://sci.esa.int/planck/
[2] http://map.gsfc.nasa.gov/
[3] http://hubblesite.org/hubble_discoveries/dark_energy/de-type_ia_supernovae.php

doi:10.1088/978-1-6817-4118-5ch7

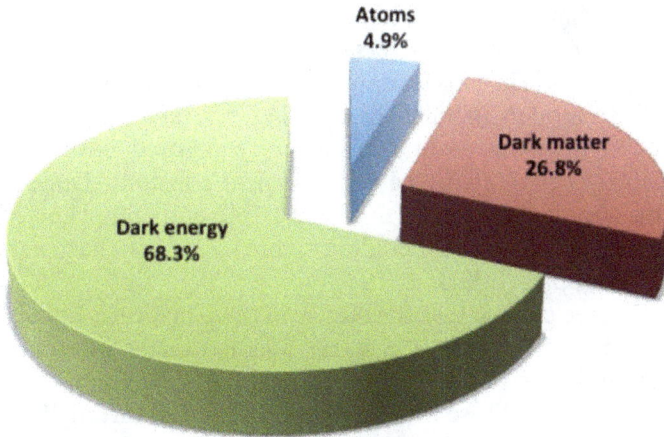

Figure 7.1. The energy budget of the Universe according to recent cosmological evidence from the Planck satellite and assuming the ΛCDM model. Atoms (or visible matter) make up just 4.9% of matter and energy in the Universe, dark matter makes up 26.8%, and dark energy makes up 68.3%.

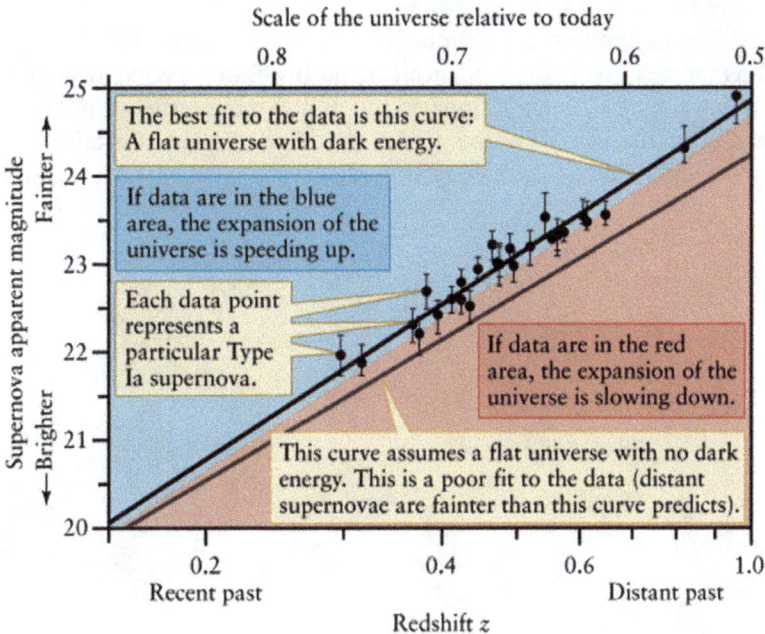

Figure 7.2. The data summarized in this illustration involve the measurement of redshifts (or recessional velocities) of distant Type 1a supernovae. The observed magnitudes are plotted as a function of their redshifts. The result shows that, based upon their redshifts, the magnitudes of the supernovae are fainter than expected. This indicates that the supernovae are further away than they should be. The solution is to invoke a universal expansion rate that has accelerated (or dark energy). Image credits: Perlmutter *et al* 1999 *Astrophys. J.* **517** 565–86, reproduced courtesy of HyperPhysics.

in laboratory experiments. Dark energy can have such a profound effect on the Universe, making up 68.3% of universal density, only because it uniformly fills otherwise empty space. The two leading models are a cosmological constant and quintessence. Both models include the common characteristic that dark energy must have negative pressure. In our currently favored cosmological model, ΛCDM, dark energy is treated as a cosmological constant. Dark energy is not the topic of this book, but, so as to avoid confusion, this short discussion is necessary. For the remaining concluding remarks in this chapter, we will now revert back to the topic of dark matter.

From figure 7.1, we can see that the matter in the Universe is dominated by dark matter. Almost 85% of the matter is dark. This is exactly what we expect from astrophysical large-scale structure formation and from extensions to the Standard Model of particle physics. These results from Planck come directly from analysis of the power spectrum of the cosmic microwave background (CMB). An image of the CMB is shown in figure 7.3 and its power spectrum in shown in figure 7.5. The Planck image of the CMB in figure 7.3 shows temperature fluctuations in the early Universe. These are the temperature fluctuations that grew into large-scale structures that we see in the Universe today, as described in chapter 2. The temperature fluctuations in the CMB are tiny, about 1 part in 10 000, with hot areas shown in red and cold areas shown in blue.

The CMB radiation and the cosmological redshift-distance relation are together regarded as the best available evidence for the Big Bang theory. Measurements of the CMB have made the inflationary Big Bang theory the standard model of cosmology. The discovery of the CMB in the mid-1960s curtailed interest in alternative models such as the steady state theory.

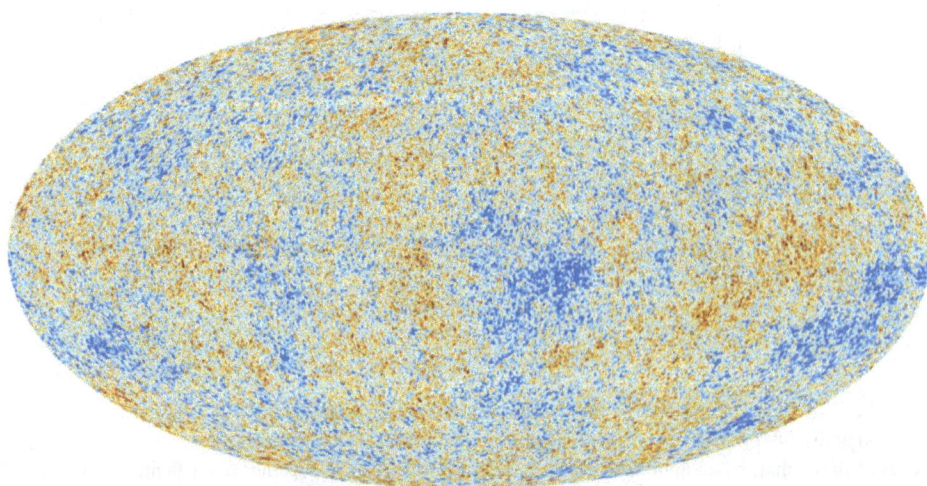

Figure 7.3. All sky image of radiation from the cosmic microwave background (CMB). Blue areas are colder than average and red areas are warmer. Temperature variations in the CMB are about ± 0.0001 K. Image courtesy of the Planck experiment/European Space Agency.

The CMB essentially confirms the Big Bang theory. In the late 1940s Alpher and Herman reasoned that if there had been a Big Bang, the expansion of the Universe would have stretched and cooled the high-energy radiation of the very early Universe into the microwave region and down to a temperature of about 5 K. They were slightly off with their estimate (it is closer to a temperature of 2.725 K), but they had exactly the right idea. They predicted the CMB. It took another 15 years for Penzias and Wilson to stumble into discovering that the microwave background was actually there, a discovery for which they won the Nobel Prize in Physics.

The CMB gives a snapshot of the Universe when, according to standard cosmology, the temperature dropped enough to allow electrons and protons to form neutral hydrogen atoms, thus making the Universe transparent to radiation. When it originated some 380 000 years after the Big Bang—this time is generally known as the 'time of last scattering' or the period of recombination or decoupling— the temperature of the Universe was about 3000 K. This corresponds to an energy of about 0.25 eV, which is much less than the 13.6 eV ionization energy of hydrogen.

Since decoupling, the temperature of the background radiation has dropped by a factor of roughly 1100 due to the expansion of the Universe. As the Universe expands, the CMB photons are redshifted, making the radiation's temperature inversely proportional to a parameter called the Universe's scale length. The temperature T_r of the CMB as a function of redshift, z, can be shown to be proportional to the temperature of the CMB as observed in the present day (2.725 K or 0.235 meV):

$$T_r = 2.725(1 + z) \tag{7.1}$$

7.1 Primary anisotropy in the CMB

The anisotropy (i.e., the temperature fluctuations) of the CMB is divided into two types: primary anisotropy, due to effects which occur at the last scattering surface and before; and secondary anisotropy, due to effects such as interactions of the background radiation with hot gas or gravitational potentials, which occur between the last scattering surface and the observer. In this chapter, we will only deal with primary anisotropies, since these can tell us something about dark matter in the Universe.

The structure of the cosmic microwave background anisotropies is principally determined by two effects: acoustic oscillations and diffusion damping (also called collisionless damping or Silk damping). The acoustic oscillations arise because of a conflict in the photon–baryon plasma in the early Universe. The pressure of the photons tends to erase anisotropies, whereas the gravitational attraction of the baryons—moving at speeds much slower than light—makes them tend to collapse to form dense halos. These two effects compete to create acoustic oscillations, which give the microwave background its characteristic peak structure. The peaks correspond, roughly, to resonances in which the photons decouple when a particular

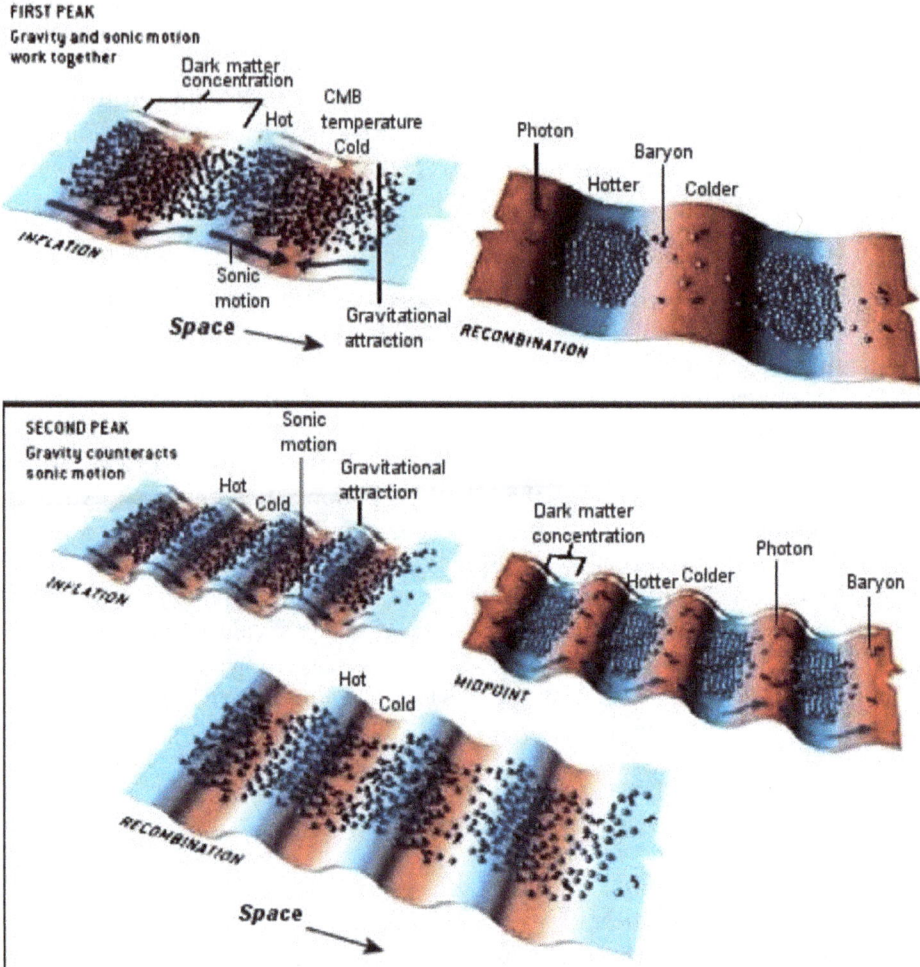

Figure 7.4. Diagrammatic explanation of the peaks in the CMB power spectrum. In the first peak of the power spectrum, gravity and sonic (i.e., acoustic) motion work together, whereas in the second peak, they work in opposite directions. For both peaks, the effects of gravity and sonic motion are shown during the inflationary period and recombination (which is the surface of last scattering). For the second peak, a midpoint is also shown.

mode is at its peak amplitude. A diagrammatic illustration of this is pictured in figure 7.4.

The peaks contain interesting physical signatures. The angular scale of the first peak determines the curvature of the Universe (but not the topology of the Universe). The next peak—ratio of the odd peaks to the even peaks—determines the reduced baryon density. The third peak can be used to get information about the dark matter density.

Figure 7.5 shows the best data we currently have for the CMB power spectrum. These data come from the European Space Agency's Planck satellite, which was

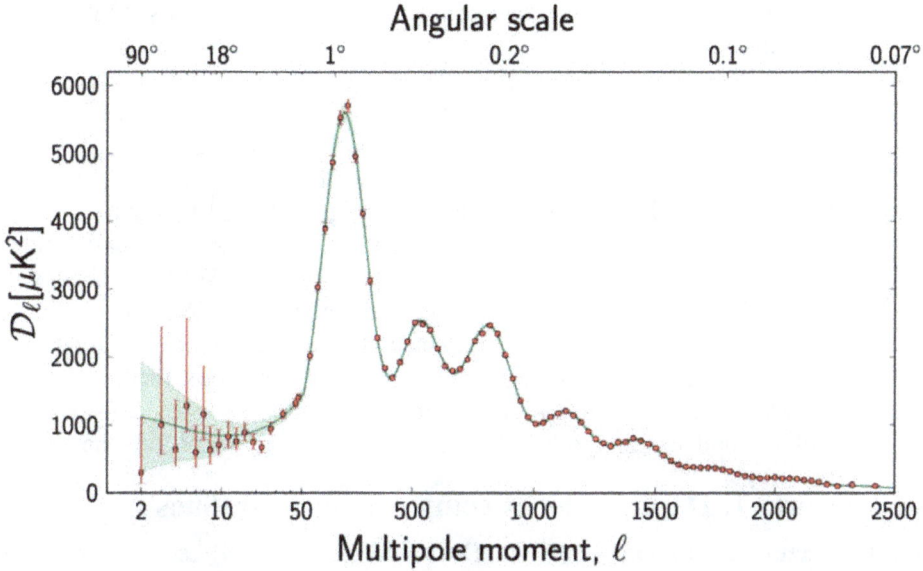

Figure 7.5. The power spectrum of the CMB radiation temperature anisotropy in terms of the angular scale (or multipole moment). The data shown comes from the Planck satellite. Also shown is a theoretical model (solid line). Image courtesy of the Planck experiment/European Space Agency.

operated from 2009 to 2013. It mapped the CMB at infrared and microwave wavelengths, with unprecedented sensitivity and angular scale. The mission substantially improved upon observations made with NASA's WMAP satellite, and it provided a major source of information relevant to the average density of normal and dark matter in the Universe.

The model line (solid black line) in figure 7.5 can be used to determine several cosmological parameters. For the purposes of this book, we will concentrate on five of these parameters. The first of these is the age of the Universe, for which Planck found $t_0 = 13.819$ billion years. Next, we have the expansion rate of the Universe, measured as the Hubble constant, for which Planck revealed $H_0 = 67.11$ km s^{-1} Mpc^{-1}. The next three parameters revealed by Planck are the energy density of dark energy Ω_Λ, regular (baryonic) matter Ω_b, and dark matter Ω_{DM}. For these three parameters, Planck revealed the following results

$$\Omega_\Lambda = 0.683$$
$$\Omega_b = 0.049 \qquad (7.2)$$
$$\Omega_{DM} = 0.268$$

which is where the numbers in the pie chart in figure 7.1 came from. In equation (7.2), each energy density is measured as a fraction of the critical density of the Universe, $\rho_c = 9.47 \times 10^{-27}$ kg m^{-1}. The total energy density is given by

$$\Omega_{tot} = \frac{\rho}{\rho_c} = \Omega_\Lambda + \Omega_b + \Omega_{DM} = 1.000. \qquad (7.3)$$

Planck and WMAP have both shown that the Universe is very close to its critical density to within a few per cent.

These results shown in equation (7.2) demonstrate that we know the average density of dark matter in the Universe extremely well. This comes not only from our measurements of the CMB, but also from measurements of the large-scale structure of the Universe as described in chapter 2. Our particle physics models also give us the correct relic densities for dark matter densities, so that the average dark matter density in the cosmos can be replicated by them. So, what are the chances of actually finding dark matter using the techniques described in chapters 5 and 6? What scenarios exist for detecting dark matter and identifying the dark matter particle? What can we conclude if the experiments outlined in chapters 5 and 6 continue to provide null results? The rest of this chapter will focus on four possible scenarios, the last of which focuses on a type of dark matter candidate (not described in the preceding chapters) that is essentially undetectable, other than through its gravitational influence.

7.2 Scenario 1: Dark matter is composed of neutralinos

Direct detection experiments (such as DarkSide or XENON) see a dark matter signal in spin-independent scattering. This result is confirmed by the LHC, which sees a missing energy signal that is followed up by precision measurements pointing to a 110 GeV neutralino. Further LHC studies constrain the neutralino's predicted thermal relic density to be identical at the per cent level with Ω_{DM}, establishing a new standard cosmology in which the dark matter is composed entirely of neutralinos back to 1 ns after the Big Bang. Supersymmetry is confirmed as the correct extension to the Standard Model of particle physics. Furthermore, direct and indirect detection rates are then used to constrain halo profiles and substructure, ushering in a new era of dark matter astrophysics.

7.3 Scenario 2: Dark matter is composed of gravitinos

The LHC discovers heavy, charged particles that are apparently stable. Together, the LHC and International Linear Collider determine that these new particles are staus, the supersymmetric partner of the tau, which is predicted by supersymmetry. Detailed follow-up studies show that, if these staus are absolutely stable, their thermal relic density is greater than the total mass of the Universe (an obvious problem)! This apparent paradox is resolved by further studies that show that staus decay on timescales of about 1 month to gravitinos. Careful studies of these decays determine that the amount of gravitinos in the Universe is exactly that required to be dark matter, providing strong quantitative evidence that dark matter is entirely in the form of gravitinos, and providing empirical support for supersymmetry, supergravity, and string theory. This scenario could also essentially explain the lower-density cores that astrophysicists observe in the centers of dwarf galaxies, i.e., small-scale structures in the Universe.

7.4 Scenario 3: Dark matter is composed of sterile neutrinos

A space-based x-ray observatory discovers a line signal. Assuming this results from decaying sterile neutrinos, the photon energy determines the mass of the sterile neutrino to be $m = 2E_\gamma/c^2$, the intensity determines the neutrino-mixing angle,

$I \propto \sin^2 \theta$, and the morphology of x-ray image determines the dark matter's spatial distribution. From the mass and radial distribution, theorists determine the free-streaming length of the sterile neutrino. This favors production from decays of the Higgs boson over production by oscillations, leading to predictions of non-standard Higgs phenomenology, which are then confirmed at the LHC. Additional information on neutrino parameters from the LHC strengthens the hypothesis of sterile neutrino dark matter, and the energy distribution of the narrow spectral line is then used to study the expansion history of the Universe, the acceleration parameter, and dark energy.

Before scenario 4 is described, it is important to stress that the above scenarios are idealized and maybe even speculative, but it also demonstrates that there are some promising detection methods in particle physics and astrophysics, which could potentially reveal the nature of dark matter in the near future. However, there is a possibility that we do not find the elusive dark matter particle using the techniques described in this book.

7.5 Scenario 4: Dark matter is 'hidden'

Despite the new experiments that have been brought online in the near future, and the LHC's attempts to find dark matter, only null results have been found. The dark matter particle has not been detected. Supersymmetry has been essentially ruled out by experimentation. All of the particles mentioned in this book have essentially been ruled out. So what prospects exist if this scenarios turns out to be true?

It is important to stress at this point, that despite theoretical progress and development of experiments, either direct or indirect, to detect dark matter, the only solid evidence we have for dark matter is from its gravitational effect. There is also strong evidence against dark matter having strong or electromagnetic interactions. It is therefore possible that dark matter has no interactions with Standard Model particles whatsoever, i.e., the dark matter is 'hidden'. Hidden dark matter has been studied, at least from a theoretical standpoint, for decades and it brings with it a huge amount of model-building freedom. This leads to a large and diverse class of particle candidates. However, this freedom comes at a cost: one typically loses connections to the central problems of particle physics discussed in chapter 4. The WIMP miracle is also gone in these hidden dark matter scenarios. Finally, because hidden dark matter candidates have no non-gravitational signals, any form of predictivity of the models is lost. This is because some non-gravitational signal is needed to identify dark matter. If dark matter is entirely hidden, we will probably never know the nature of dark matter. We will just know that dark matter exists (because of its gravitational signal), and we would have ruled out every possible particle that has some kind of non-gravitational signal.

7.6 Concluding remarks

The favored form of dark matter is the neutralino, a WIMP, predicted by super-symmetry, simply because neutralino (and WIMP models in general) produce the correct relic density for dark matter, a coincidence termed the WIMP miracle. As yet, LHC has found no evidence in favor of supersymmetry, but it is about to start searching for new particles at higher energies, so there is still hope. Other alternatives to neutralinos, and even WIMPs, have been proposed, but these alternative models are not as elegant because they do not naturally reproduce the required relic density.

As for the ultimate nature of dark matter (and even dark energy), at present we can only infer its existence from gaps in current cosmological theory and supporting data. It is entirely probable that the ultimate answer to these puzzles lies outside the realm of dualistic perception, that of the observer and the observed, a central thesis of quantum theory. Resolving this enigma will require delving into the nature of emptiness itself, that the Universe has indeed arisen out of emptiness and thus emptiness is its inherent nature. The theoretical implications of this are profound. Our best models are still unable to address the nature of a singularity, a point of infinite density (a model, for example, of a black hole). At a singularity, the laws of physics break down and we are left dangling on string theory and membranes as our best hope of resolving what is arguably beyond the human mind's ability to grasp. Great experiments and instruments such as these will continue to improve our understanding of the physical Universe but these basic questions may not be resolved until such time as the mathematical and theoretical nature of emptiness, as distinct from nothingness or voidness, is fully embraced by modern science. Until then the Cheshire cat will grin knowingly at us from his sublime perch.

Suggested further reading

Komatsu E *et al* 2011 *Astrophys. J. Suppl. Ser.* **192** 18–47
Penzias A A and Wilson R W 1965 *Astrophys. J.* **142** 419–21
Perlmutter S *et al* 1999 *Astrophys. J.* **517** 565–86
Planck Collaboration 2014 *Astron. Astrophys.* **571** A15
Planck Collaboration 2014 *Astron. Astrophys.* **571** A16
Planck Collaboration 2014 *Astron. Astrophys.* **571** A24
Riess A G *et al* 1998 *Astrophys. J.* **116** 1009–38
Schmidt B P *et al* 1998 *Astrophys. J.* **507** 46–63